zero

zero

The Biography of a Dangerous Idea

[C H A R L E S   S E I F E]

Drawings by Matt Zimet

Viking

VIKING
Published by the Penguin Group
Penguin Putnam Inc., 375 Hudson Street,
New York, New York 10014, U.S.A.
Penguin Books Ltd, 27 Wrights Lane,
London W8 5TZ, England
Penguin Books Australia Ltd, Ringwood,
Victoria, Australia
Penguin Books Canada Ltd, 10 Alcorn Avenue,
Toronto, Ontario, Canada M4V 3B2
Penguin Books (N.Z.) Ltd, 182–190 Wairau Road,
Auckland 10, New Zealand

Penguin Books Ltd, Registered Offices:
Harmondsworth, Middlesex, England

First published in 2000 by Viking Penguin,
a member of Penguin Putnam Inc.

3   5   7   9   10   8   6   4   2

Illustration credits
Pages 30, 31, 69, 85: Courtesy of The Library of Congress
Page 65: The Nelson-Atkins Museum of Art, Kansas City, Missouri (Purchase:
Nelson Trust)

All other drawings by Matt Zimet.

A portion of Chapter 7 appeared in different form as "Running on Empty,"
New Scientist, April 25, 1998.

LIBRARY OF CONGRESS CATALOGING-IN-PUBLICATION DATA
Seife, Charles.
Zero : the biography of a number / Charles Seife.
p.   cm.
Includes bibliographical references.
ISBN 0-670-88457-X
1. Zero (The number)   I. Title.
QA141.S45   2000
511.2'11 — dc21   99–36693
This book is printed on acid-free paper.

Printed in the United States of America
Set in Cochin
Designed by Jaye Zimet

Contents

zero

Chapter **0**
Null and Void

Z ero hit the USS *Yorktown* like a torpedo.

On September 21, 1997, while cruising off the coast of Virginia, the billion-dollar missile cruiser shuddered to a halt. *Yorktown* was dead in the water.

Warships are designed to withstand the strike of a torpedo or the blast of a mine. Though it was armored against weapons, nobody had thought to defend the *Yorktown* from zero. It was a grave mistake.

The *Yorktown*'s computers had just received new software that was controlling the engines. Unfortunately, nobody had spotted the time bomb lurking in the code, a zero that engineers were supposed to remove while installing the software. But for one reason or another, the zero was overlooked, and it stayed hidden in the code. Hidden, that is, until the software called it into memory—and choked.

When the *Yorktown*'s computer system tried to divide by zero, 80,000 horsepower instantly became worthless. It took nearly three hours to attach emergency controls to the engines, and the *Yorktown* then limped into port. Engineers spent two days getting rid of the zero, repairing the engines, and putting the *Yorktown* back into fighting trim.

No other number can do such damage. Computer failures like the one that struck the *Yorktown* are just a faint shadow of the power of zero. Cultures girded themselves against zero, and philosophies crumbled under its influence, for zero is different from the other numbers. It provides a glimpse of the ineffable and the infinite. This is why it has been feared and hated—and outlawed.

This is the story of zero, from its birth in ancient times to its growth and nourishment in the East, its struggle for acceptance in Europe, its ascendance in the West, and its ever-present threat to modern physics. It is the story of the people who battled over the meaning of the mysterious number—the scholars and mystics, the scientists and clergymen—who each tried to understand zero. It is the story of the Western world's attempts to shield itself unsuccessfully (and sometimes violently) from an Eastern idea. And it is a history of the paradoxes posed by an innocent-looking number, rattling even this century's brightest minds and threatening to unravel the whole framework of scientific thought.

Zero is powerful because it is infinity's twin. They are equal and opposite, yin and yang. They are equally paradoxical and troubling. The biggest questions in science and religion are about nothingness and eternity, the void and the infinite, zero and infinity. The clashes over zero were the battles that shook the foundations of philosophy, of science, of mathematics, and of religion. Underneath every revolution lay a zero—and an infinity.

Zero was at the heart of the battle between East and West. Zero was at the center of the struggle between religion and science. Zero became the language of nature and the most important tool in mathematics. And the most profound problems in physics—the dark core of a black hole and the brilliant flash of the big bang—are struggles to defeat zero.

Yet through all its history, despite the rejection and the exile, zero has always defeated those who opposed it. Humanity could never force zero to fit its philosophies. Instead, zero shaped humanity's view of the universe—and of God.

Chapter **1**
Nothing Doing

[THE ORIGIN OF ZERO]

There was neither non-existence nor existence then; there was neither the realm of space nor the sky which is beyond. What stirred? Where?

—THE RIG VEDA

The story of zero is an ancient one. Its roots stretch back to the dawn of mathematics, in the time thousands of years before the first civilization, long before humans could read and write. But as natural as zero seems to us today, for ancient peoples zero was a foreign—and frightening—idea. An Eastern concept, born in the Fertile Crescent a few centuries before the birth of Christ, zero not only evoked images of a primal void, it also had dangerous mathematical properties. Within zero there is the power to shatter the framework of logic.

The beginnings of mathematical thought were found in the desire to count sheep and in the need to keep track of property and of the passage of time. None of these tasks requires zero; civilizations functioned perfectly well for millennia before its discovery. Indeed, zero was so abhorrent to some cultures that they chose to live without it.

Life without Zero

The point about zero is that we do not need to use it in the operations of daily life. No one goes out to buy zero fish. It is in a way the most civilized of all the cardinals, and its use is only forced on us by the needs of cultivated modes of thought.

—ALFRED NORTH WHITEHEAD

It's difficult for a modern person to imagine a life without zero, just as it's hard to imagine life without the number seven or the number 31. However, there was a time where there was no zero—just as there was no seven and 31. It was before the beginning of history, so paleontologists have had to piece together the tale of the birth of mathematics from bits of stone and bone. From these fragments, researchers discovered that Stone Age mathematicians were a bit more rugged than modern ones. Instead of blackboards, they used wolves.

A key clue to the nature of Stone Age mathematics was unearthed in the late 1930s when archaeologist Karl Absolom, sifting through Czechoslovakian dirt, uncovered a 30,000-year-old wolf bone with a series of notches carved into it. Nobody knows whether Gog the caveman had used the bone to count the deer he killed, the paintings he drew, or the days he had gone without a bath, but it is pretty clear that early humans were counting something.

A wolf bone was the Stone Age equivalent of a supercomputer. Gog's ancestors couldn't even count up to two, and they certainly did not need zero. In the very beginning of mathematics, it seems that people could only distinguish between *one* and *many*. A caveman owned one spearhead or many spearheads; he had eaten one crushed lizard or many crushed lizards. There was no way to express any quantities other than one and many. Over time, primitive languages evolved to distinguish between *one, two,* and *many,* and eventually *one, two, three, many,* but didn't have terms for higher numbers. Some languages still have this shortcoming. The Siriona Indians of Bolivia and the Brazilian Yanoama people don't have words for anything larger than three; instead, these two tribes use the words for "many" or "much."

Thanks to the very nature of numbers—they can be added together to create new ones—the number system didn't stop at three. After a while, clever tribesmen began to string number-words in a row to yield more numbers. The languages currently used by the Bacairi and the Bororo peoples of Brazil show this process in action; they have number systems that go "one," "two," "two and one," "two and two," "two and two and one," and so forth. These people count by twos. Mathematicians call this a *binary* system.

Few people count by twos like the Bacairi and Bororo. The old wolf bone seems to be more typical of ancient counting systems. Gog's wolf bone had 55 little notches in it, arranged into groups of five; there was a second notch after the first 25 marks. It looks suspiciously as if Gog was counting by fives, and then tallied groups in bunches of five. This makes a lot of sense. It is a lot faster to tally the number of marks in groups than it is to count them one by one. Modern mathematicians would say that Gog, the wolf carver, used a five-based or *quinary* counting system.

But why five? Deep down, it's an arbitrary decision. If Gog put his tallies in groups of four, and counted in groups of four and 16, his number system would have worked just as well, as would groups of six and 36. The groupings don't affect the number of marks on the bone; they only affect the way that Gog tallies them up in the end—and he will always get the same answer no matter how he counts them. However, Gog preferred to count in groups of five rather than four, and people all over the world shared Gog's preference. It was an accident of nature that gave humans five fingers on each hand, and because of this accident, five seemed to be a favorite base system across many cultures. The early Greeks, for instance, used the word "fiving" to describe the process of tallying.

Even in the South American binary counting schemes, linguists see the beginnings of a quinary system. A different phrase in Bororo for "two and two and one" is "this is my hand all together." Apparently, ancient peoples liked to count with their body parts, and five (a hand), ten (both hands), and twenty (both hands and both feet) were the favorites. In English, eleven and twelve seem to be derived from "one over [ten]" and "two over [ten]," while thirteen, fourteen, fifteen, and so on are contractions of "three and ten," "four and ten," and "five and ten." From this, linguists conclude that ten was the basic unit in the Germanic protolanguages that English came from, and thus those people used a base-10 number system. On the other hand, in French, eighty is *quatre-vingts* (four twenties), and ninety is *quatre-vingt-dix* (four twenties and ten). This may mean that the people who lived in what is now France used a base-20 or *vigesimal* number system. Numbers like seven and 31 belonged to all of these systems, quinary, decimal, and vigesimal alike. However, none of these systems had a name for zero. The concept simply did not exist.

You never need to keep track of zero sheep or tally your zero children. Instead of "We have zero bananas," the grocer says, "We have no bananas." You don't have to have a number to express the lack of something, and it didn't occur to anybody to assign a symbol to the absence of objects. This is why people got along without zero for so long. It simply wasn't needed. Zero just never came up.

In fact, knowing about numbers at all was quite an ability in prehistoric times. Simply being able to count was considered a talent as mystical and arcane as casting spells and calling the gods by name. In the Egyptian Book of the Dead, when a dead soul is challenged by Aqen, the ferryman who conveys departed spirits across a river in the netherworld,

Aqen refuses to allow anyone aboard "who does not know the number of his fingers." The soul must then recite a counting rhyme to tally his fingers, satisfying the ferryman. (The Greek ferryman, on the other hand, wanted money, which was stowed under the dead person's tongue.)

Though counting abilities were rare in the ancient world, numbers and the fundamentals of counting always developed before writing and reading. When early civilizations started pressing reeds to clay tablets, carving figures in stone, and daubing ink on parchment and on papyrus, number systems had already been well-established. Transcribing the oral number system into written form was a simple task: people just needed to figure out a coding method whereby scribes could set the numbers down in a more permanent form. (Some societies even found a way to do this before they discovered writing. The illiterate Incas, for one, used the *quipu*, a string of colored, knotted cords, to record calculations.)

The first scribes wrote down numbers in a way that matched their base system, and predictably, did it in the most concise way they could think of. Society had progressed since the time of Gog. Instead of making little groups of marks over and over, the scribes created symbols for each type of grouping; in a quinary system, a scribe might make a certain mark for one, a different symbol for a group of five, yet another mark for a group of 25, and so forth.

The Egyptians did just that. More than 5,000 years ago, before the time of the pyramids, the ancient Egyptians designed a system for transcribing their decimal system, where pictures stood for numbers. A single vertical mark represented a unit, while a heel bone represented 10, a swirly snare stood for 100, and so on. To write down a number with this scheme, all an Egyptian scribe had to do was record groups of these symbols. Instead of having to write down 123 tick marks to denote the number "one hundred and twenty-three," the scribe wrote six symbols: one snare, two heels, and three vertical marks. It was the typical way of doing mathematics in antiquity. And like most other civilizations Egypt did not have—or need—a zero.

Yet the ancient Egyptians were quite sophisticated mathematicians. They were master astronomers and timekeepers, which meant that they had to use advanced math, thanks to the wandering nature of the calendar.

Creating a stable calendar was a problem for most ancient peoples, because they generally started out with a lunar calendar: the length of a month was the time between successive full moons. It was a natural choice; the waxing and waning of the moon in the heavens was hard to overlook, and it offered a convenient way of marking periodic cycles of time. But the lunar month is between 29 and 30 days long. No matter how you arrange it, 12 lunar months only add up to about 354 days—roughly 11 short of the solar year's length. Thirteen lunar months yield roughly 19 days too many. Since it is the solar year, not the lunar year, that determines the time for harvest and planting, the seasons seem to drift when you reckon by an uncorrected lunar year.

Correcting the lunar calendar is a complicated undertaking. A number of modern-day nations, like Israel and Saudi Arabia, still use a modified lunar calendar, but 6,000 years ago the Egyptians came up with a better system. Their method was a much simpler way of keeping track of the passage of the days, producing a calendar that stayed in sync with the seasons for many years. Instead of using the moon to keep track of the passage of time, the Egyptians used the sun, just as most nations do today.

The Egyptian calendar had 12 months, like the lunar one, but each month was 30 days long. (Being base-10 sort of people, their week, the *decade*, was 10 days long.) At the end of the year, there were an extra five days, bringing the total up to 365. This calendar was the ancestor of our own calendar; the Egyptian system was adopted by Greece and then by Rome, where it was modified by adding leap years, and then became the standard calendar of the Western world. However, since the Egyptians, the Greeks, and the Romans did not have zero, the Western calendar does not have any zeros—an oversight that would cause problems millennia later.

The Egyptians' innovation of the solar calendar was a breakthrough, but they made an even more important mark on history: the invention of the art of geometry. Even without a zero, the Egyptians had quickly become masters of mathematics. They had to, thanks to an angry river. Every year the Nile would overflow its banks and flood the delta. The good news was that the flooding deposited rich, alluvial silt all over the fields, making the Nile delta the richest farmland in the ancient world. The bad news was that the river destroyed many of the boundary markers, erasing all of the landmarks that told farmers which land was theirs to cultivate. (The Egyptians took property rights very seriously. In the Egyptian Book of the Dead, a newly deceased person must swear to the gods that he hasn't cheated his neighbor by stealing his land. It was a sin punishable by having his heart fed to a horrible beast called the devourer. In Egypt, filching your neighbor's land was considered as grave an offense as breaking an oath, murdering somebody, or masturbating in a temple.)

The ancient pharaohs assigned surveyors to assess the damage and reset the boundary markers, and thus geometry was born. These surveyors, or rope stretchers (named for their measuring devices and knotted ropes designed to mark right angles), eventually learned to determine the areas of plots of land by dividing them into rectangles and triangles. The Egyptians also learned how to measure the volumes of objects—like pyramids. Egyptian mathematics was famed throughout the Mediterranean, and it is likely that the early Greek mathematicians, masters of geometry like Thales and Pythagoras, studied in Egypt. Yet despite the Egyptians' brilliant geometric work, zero was nowhere to be found within Egypt.

This was, in part, because the Egyptians were of a practical bent. They never progressed beyond measuring volumes and counting days and hours. Mathematics wasn't used for anything impractical, except their system of astrology. As a result, their best mathematicians were unable to use the principles of geometry for anything unrelated to real world problems — they did not take their system of mathematics and turn it into an abstract system of logic. They were also not inclined to put math into their philosophy. The Greeks were different; they embraced the abstract and the philosophical, and brought mathematics to its highest point in ancient times. Yet it was not the Greeks who discovered zero. Zero came from the East, not the West.

The Birth of Zero

In the history of culture the discovery of zero will always stand out as one of the greatest single achievements of the human race.

— TOBIAS DANZIG, *NUMBER:
THE LANGUAGE OF SCIENCE*

The Greeks understood mathematics better than the Egyptians did; once they mastered the Egyptian art of geometry, Greek mathematicians quickly surpassed their teachers.

At first the Greek system of numbers was quite similar to the Egyptians'. Greeks also had a base-10 style of counting, and there was very little difference in the ways the two cultures wrote down their numbers. Instead of using pictures to represent numbers as the Egyptians did, the Greeks used letters. H (eta) stood for *hekaton*: 100. M (mu) stood for *myriori*: 10,000 — the myriad, the biggest grouping in the Greek system. They also had a symbol for five, indicating a mixed quinary-decimal system, but overall the Greek and Egyptian systems of writing numbers were almost identical — for a time. Unlike the Egyptians, the Greeks outgrew this primitive way of writing numbers and developed a more sophisticated system.

Instead of using two strokes to represent 2, or three Hs to represent 300 as the Egyptian style of counting did, a newer Greek system of writing, appearing before 500 BC, had distinct letters for 2, 3, 300, and many other numbers (Figure 1). In this way the Greeks avoided repeated letters. For instance, writing the number 87 in the Egyptian system would require 15 symbols: eight heels and seven vertical marks. The new Greek system would need only two symbols: π for 80, and ζ for 7. (The Roman system, which supplanted Greek numbers, was a step backward toward the less sophisticated Egyptian system. The Roman 87, LXXXVII, requires seven symbols, with several repeats.)

Though the Greek number system was more sophisticated than the Egyptian system, it was not the most advanced way of writing numbers in the ancient world. That title was held by another Eastern invention: the Babylonian style of counting. And thanks to this system, zero finally appeared in the East, in the Fertile Crescent of present-day Iraq.

At first glance the Babylonian system seems perverse. For one thing the system is *sexagesimal* — based on the number 60. This is an odd-looking choice, especially since most human societies chose 5, 10, or 20 as their base number. Also, the Babylonians used only two marks to represent their numbers: a wedge that represented 1 and a double wedge that represented 10. Groups of these marks, arranged in clumps

Figure 1: Numerals of different cultures

	1	2	3	4	10	20	30	100	200	123
MODERN	1	2	3	4	10	20	30	100	200	123
EGYPTIAN	I	II	III	IIII	∩	∩∩	∩∩∩	ℓ	ℓℓ	ℓ∩∩III
GREEK (OLD STYLE)	I	II	III	IIII	Δ	ΔΔ	ΔΔΔ	H	HH	HΔΔIII
GREEK (NEW STYLE)	α	β	γ	δ	ι	κ	λ	ρ	σ	ρκγ
ROMAN	I	II	III	IV	X	XX	XXX	C	CC	CXXIII
HEBREW	א	ב	ג	ד	ˋ	כ	ל	ק	ר	קכג
MAYAN	·	··	···	····	=	·̲̲	·̲·̲	̲	̳	·̈·

that summed to 59 or less, were the basic symbols of the counting system, just as the Greek system was based on letters and the Egyptian system was based on pictures. But the really odd feature of the Babylonian system was that, instead of having a different symbol for each number like the Egyptian and Greek systems, each Babylonian symbol could represent a multitude of different numbers. A single wedge, for instance, could stand for 1; 60; 3,600; or countless others.

As strange as this system seems to modern eyes, it made perfect sense to ancient peoples. It was the Bronze Age equivalent of computer code. The Babylonians, like many different cultures, had invented machines that helped them count. The most famous was the abacus. Known as the *soroban* in Japan, the *suan-pan* in China, the *s'choty* in Russia, the *coulba* in Turkey, the *choreb* in Armenia, and by a variety of other names in different cultures, the abacus relies upon sliding stones to keep track of amounts. (The words *calculate, calculus,* and *calcium* all come from the Latin word for pebble: calculus.)

Adding numbers on an abacus is as simple as moving the stones up and down. Stones in different columns have different values, and by manipulating them a skilled user can add large numbers with great speed. When a calculation is complete, all the user has to do is look at the final position of the stones and translate that into a number—a pretty straightforward operation.

The Babylonian system of numbering was like an abacus inscribed symbolically onto a clay tablet. Each grouping of symbols represented a certain number of stones that had been moved on the abacus, and like each column of the abacus, each grouping had a different value, depending on its position. In this way the Babylonian system was not so different from the system we use today. Each 1 in the number 111 stands for a different value; from right to left, they stand for "one," "ten," and "one hundred," respectively. Similarly, the symbol Y in YYY stood for "one," "sixty," or "thirty-six hundred" in the three different positions. It was just like an abacus, except for one problem. How would a Babylonian write the number 60? The number 1 was easy to write: Y. Unfortunately, 60 was also written as Y; the only difference was that Y was in the second position rather than the first. With the abacus it's easy to tell which number is represented. A single stone in the first column is easy to distinguish from a single stone in the second column. The same isn't true for writing. The Babylonians had no way to denote which column a written symbol was in; Y could represent 1, 60, or 3,600. It got worse when they mixed numbers. The symbol YY could mean 61; 3,601; 3,660; or even greater values.

Zero was the solution to the problem. By around 300 BC the Babylonians had started using two slanted wedges, �, to represent an empty space, an empty column on the abacus. This *placeholder* mark made it easy to tell which position a symbol was in. Before the advent of zero, YY could be interpreted as 61 or 3,601. But with zero, YY meant 61; 3,601 was written as Y ⋗ Y (Figure 2). Zero was born out of the need to give any given sequence of Babylonian digits a unique, permanent meaning.

Though zero was useful, it was only a placeholder. It was merely a symbol for a blank place in the abacus, a column where all the stones were at the bottom. It did little more than make sure digits fell in the right places; it didn't really have a numerical value of its own. After all, 000,002,148 means exactly the same thing as 2,148. A zero in a string of digits takes its meaning from some other digit to its left. On its own, it meant . . . nothing. Zero was a digit, not a number. It had no value.

Figure 2: Babylonian numbers

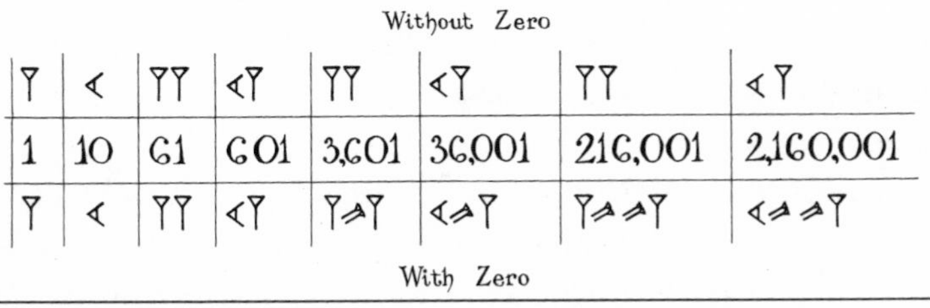

Without Zero

Y	⟨	YY	⟨Y	YY	⟨Y	YY	⟨ Y
1	10	61	601	3,601	36,001	216,001	2,160,001
Y	⟨	YY	⟨Y	Y⋗Y	⟨⋗Y	Y⋗⋗Y	⟨⋗ ⋗Y

With Zero

A number's value comes from its place on the number line—from its position compared with other numbers. For instance, the number two comes before the number three and after the number one; nowhere else makes any sense. However, the 0 mark didn't have a spot on the number line at first. It was just a symbol; it didn't have a place in the hierarchy of numbers. Even today, we sometimes treat zero as a nonnumber even though we all know that zero has a numerical value of its own, using the digit 0 as a placeholder without connecting it to the number zero. Look at a telephone or the top of a computer keyboard. The 0 comes after the 9, not before the 1 where it belongs. It doesn't matter where the placeholder 0 sits; it can be anywhere in the number sequence. But nowadays everybody knows that zero can't really sit anywhere on the number line, because it has a definite numerical value of its own. It is the number that separates the positive numbers from the negative numbers. It is an even number, and it is the integer that precedes one. Zero must sit in its rightful place on the number line, before one and after negative one. Nowhere else makes any sense. Yet zero sits at the end of the computer and at the bottom of the telephone because we always start counting with one.

One seems like the appropriate place to start counting, but doing so forces us to put zero in an unnatural place. To other cultures, like the Mayan people of Mexico and Central America, starting with one didn't seem like the rational thing to do. In fact, the Mayans had a number system—and a calendar—that made more sense than ours does. Like the Babylonians, the Mayans had a *place-value* system of digits and places. The only real difference was that instead of basing their numbers on 60 as the Babylonians did, the Mayans had a vigesimal, base-20 system that had the remnants of an earlier base-10 system in it. And like the Babylonians, they needed a zero to keep track of what each digit meant. Just to make things interesting, the Mayans had two types of digits. The simple type was based on dots and lines, while the complicated type was based on glyphs—grotesque faces. To a modern eye, Mayan glyph writing is about as alien-looking as you can get (Figure 3).

Like the Egyptians, the Mayans also had an excellent solar calendar. Because their system of counting was based on the number 20, the Mayans naturally divided their year into 18 months of 20 days each, totaling 360 days. A special period of five days at the end, called Uayeb, brought the count to 365. Unlike the Egyptians, though, the Mayans had a zero in their counting system, so they did the obvious thing: they started numbering days with the number zero. The first day of the month of Zip, for example, was usually called the "in-

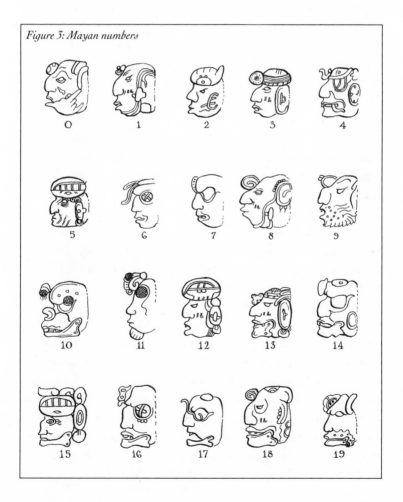

Figure 3: Mayan numbers

stallation" or "seating" of Zip. The next day was 1 Zip, the following day was 2 Zip, and so forth, until they reached 19 Zip. The next day was the seating of Zotz'—0 Zotz' followed by 1 Zotz' and so forth. Each month had 20 days, numbered 0 through 19, not numbered 1 through 20 as we do today. (The Mayan calendar was wonderfully complicated. Along with this solar calendar, there was a ritual calendar that had 20 weeks, each of 13 days. Combined with the solar year, this created a *calendar round* that had a different name for every day in a 52-year cycle.)

The Mayan system made more sense than the Western system does. Since the Western calendar was created at a time when there was no zero, we never see a day zero, or a year zero. This apparently insignificant omission caused a great deal of trouble; it kindled the controversy over the start of the millennium. The Mayans would never have argued about whether 2000 or 2001 was the first year in the twenty-first century. But it was not the Mayans who formed our calendar; it was the Egyptians and, later, the Romans. For this reason, we are stuck with a troublesome, zero-free calendar.

The Egyptian civilization's lack of zero was bad for the calendar and bad for the future of Western mathematics. In fact, Egyptian civilization was bad for math in more ways than one; it was not just the absence of a zero that caused future difficulties. The Egyptians had an extremely cumbersome way of handling fractions. They didn't think of 3/4 as a ratio of three to four as we do today; they saw it as the sum of 1/2 and 1/4. With the sole exception of 2/3, all Egyptian fractions were written as a sum of numbers in the form of $1/n$ (where n is a counting number)—the so-called unit fractions. Long chains of these unit fractions made ratios extremely difficult to handle in the Egyptian (and Greek) number systems.

Zero makes this cumbersome system obsolete. In the Babylonian system—with zero in it—it's easy to write fractions. Just as we can write 0.5 for 1/2 and 0.75 for 3/4, the Babylonians used the numbers 0;30 for 1/2 and 0;45 for 3/4. (In fact, the Babylonian base-60 system is even better suited to writing down fractions than our modern-day base-10 system.)

Unfortunately, the Greeks and Romans hated zero so much that they clung to their own Egyptian-like notation rather than convert to the Babylonian system, even though the Babylonian system was easier to use. For intricate calculations, like those needed to create astronomical tables, the Greek system was so cumbersome that the mathematicians converted the unit fractions to the Babylonian sexagesimal system, did the calculations, and then translated the answers back into the Greek style. They could have saved many time-consuming steps. (We all know how fun it is to convert fractions back and forth!) However, the Greeks so despised zero that they refused to admit it into their writings, even though they saw how useful it was. The reason: zero was dangerous.

The Fearsome Properties of Nothing

In earliest times   did Ymir live:

was nor sea nor land   nor salty waves,

neither earth was there   nor upper heaven,

but a gaping nothing,   and green things nowhere.

— THE ELDER EDDA

It is hard to imagine being afraid of a number. Yet zero was inexorably linked with the void—with nothing. There was a primal fear of void and chaos. There was also a fear of zero.

Most ancient peoples believed that only emptiness and chaos were present before the universe came to be. The Greeks claimed that at first Darkness was the mother of all things, and from Darkness sprang Chaos. Darkness and Chaos then spawned the rest of creation. The Hebrew creation myths say that the earth was chaotic and void before God showered it with light and formed its features. (The Hebrew phrase is *tohu v'bohu*. Robert Graves linked these *tohu* to Tehomot, a primal Semitic dragon that was present at the birth of the universe and whose body became the sky and earth. *Bohu* was linked to Behomot, the famed Behemoth monster of Hebrew legend.) The older Hindu tradition tells of a creator who churns the butter of chaos into the earth, and the Norse myth tells a tale of an open void that gets covered with ice, and from the chaos caused by the mingling of fire and ice was born the primal Giant. Emptiness and disorder were the primeval, natural state of the cosmos, and there was always a nagging fear that at the end of time, disorder and void would reign once more. Zero represented that void.

But the fear of zero went deeper than unease about the void. To the ancients, zero's mathematical properties were inexplicable, as shrouded in mystery as the birth of the universe. This is because zero is different from the other numbers. Unlike the other digits in the Babylonian system, zero never was allowed to stand alone—for good reason. A lone zero always misbehaves. At the very least it does not behave the way other numbers do.

Add a number to itself and it changes. One and one is not one—it's two. Two and two is four. But zero and zero is zero. This violates a basic principle of numbers called the axiom of Archimedes, which says that if you add something to itself enough times, it will exceed any other number in magnitude. (The axiom of Archimedes was phrased in terms of areas; a number was viewed as the difference of two unequal areas.) Zero refuses to get bigger. It also refuses to make any other number bigger. Add two and zero and you get two; it is as if you never bothered to add the numbers in the first place. The same thing happens with subtraction. Take zero away from two and you get two. Zero has no substance. Yet this substanceless number threatens to undermine the simplest operations in mathematics, like multiplication and division.

In the realm of numbers, multiplication is a stretch— literally. Imagine that the number line is a rubber band with tick marks on it (Figure 4). Multiplying by two can be

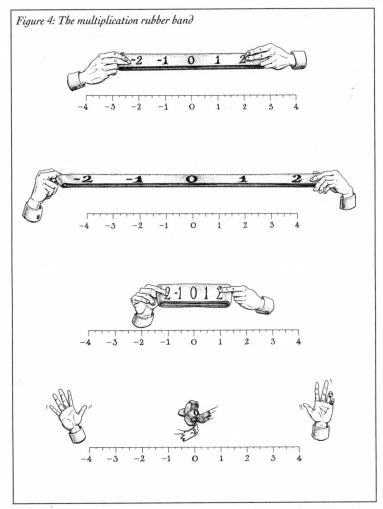

Figure 4: The multiplication rubber band

thought of as stretching out the rubber band by a factor of two: the tick mark that was at one is now at two; the tick mark that was at three is now at six. Likewise, multiplying by one-half is like relaxing the rubber band a bit: the tick mark at two is now at one, and the tick mark at three winds up at one and a half. But what happens when you multiply by zero?

Anything times zero is zero, so all the tick marks are at zero.

The rubber band has broken. The whole number line has collapsed.

Unfortunately, there is no way to get around this unpleasant fact. Zero times anything must be zero; it's a property of our number system. For everyday numbers to make sense, they have to have something called the *distributive property*, which is best seen through an example. Imagine that a toy store sells balls in groups of two and blocks in groups of three. The neighboring toy store sells a combination pack with two balls and three blocks in it. One bag of balls and one bag of blocks is the same thing as one combination package from the neighboring store. To be consistent, buying seven bags of balls and seven bags of blocks from one toy store has to be the same thing as buying seven combination packs from the neighboring shop. This is the distributive property. Mathematically speaking, we say that $7 \times 2 + 7 \times 3 = 7 \times (2 + 3)$. Everything comes out right.

Apply this property to zero and something strange happens. We know that $0 + 0 = 0$, so a number multiplied by zero is the same thing as multiplying by $(0 + 0)$. Taking two as an example, $2 \times 0 = 2 \times (0 + 0)$, but by the distributive property we know that $2 \times (0 + 0)$ is the same thing as $2 \times 0 + 2 \times 0$. But this means $2 \times 0 = 2 \times 0 + 2 \times 0$. Whatever 2×0 is, when you add it to itself, it stays the same. This seems a lot like zero. In fact, that is just what it is. Subtract 2×0 from each side of the equation and we see that $0 = 2 \times 0$. Thus, no matter what you do, multiplying a number by zero gives you zero. This troublesome number crushes the number line into a point. But as annoying as this property was, the true power of zero becomes apparent with division, not multiplication.

Just as multiplying by a number stretches the number line, dividing shrinks it. Multiply by two and you stretch the number line by a factor of two; divide by two and you relax the rubber band by a factor of two, undoing the multiplication. Divide by a number and you undo the multiplication: a tick mark that had been stretched to a new place on the number line resumes its original position.

We saw what happened when you multiply a number by zero: the number line is destroyed. Division by zero should be the opposite of multiplying by zero. It should undo the destruction of the number line. Unfortunately, this isn't quite what happens.

In the previous example we saw that 2×0 is 0. Thus to undo the multiplication, we have to assume that $(2 \times 0)/0$ will get us back to 2. Likewise, $(3 \times 0)/0$ should get us back to 3, and $(4 \times 0)/0$ should equal 4. But 2×0 and 3×0 and 4×0 each equal zero, as we saw—so $(2 \times 0)/0$ equals $0/0$, as do $(3 \times 0)/0$ and $(4 \times 0)/0$. Alas, this means that $0/0$ equals 2, but it also equals 3, and it also equals 4. This just doesn't make any sense.

Strange things also happen when we look at $1/0$ in a different way. Multiplication by zero should undo division by zero, so $1/0 \times 0$ should equal 1. However, we saw that anything multiplied by zero equals zero! There is no such number that, when multiplied by zero, yields one—at least no number that we've met.

Worst of all, if you wantonly divide by zero, you can destroy the entire foundation of logic and mathematics. Dividing by zero once—just one time—allows you to prove, mathematically, anything at all in the universe. You can prove that $1 + 1 = 42$, and from there you can prove that J. Edgar Hoover was a space alien, that William Shakespeare came from Uzbekistan, or even that the sky is polka-dotted. (See appendix A for a proof that Winston Churchill was a carrot.)

Multiplying by zero collapses the number line. But dividing by zero destroys the entire framework of mathematics.

There is a lot of power in this simple number. It was to become the most important tool in mathematics. But thanks to the odd mathematical and philosophical properties of zero, it would clash with the fundamental philosophy of the West.

Chapter 2
Nothing Comes of Nothing

[THE WEST REJECTS ZERO]

Nothing can be created from nothing.

— LUCRETIUS, *DE RERUM NATURA*

Zero clashed with one of the central tenets of Western phi-losophy, a dictum whose roots were in the number-philosophy of Pythagoras and whose importance came from the paradoxes of Zeno. The whole Greek universe rested upon this pillar: there is no void.

The Greek universe, created by Pythagoras, Aristotle, and Ptolemy, survived long after the collapse of Greek civilization. In that universe there is no such thing as nothing. There is no zero. Because of this, the West could not accept zero for nearly two millennia. The consequences were dire. Zero's absence would stunt the growth of mathematics, stifle innovation in science, and, incidentally, make a mess of the calendar. Before they could accept zero, philosophers in the West would have to destroy their universe.

The Origin of Greek Number-Philosophy

*In the beginning, there was the ratio, and the ratio was
with God, and the ratio was God.**

— JOHN 1:1

The Egyptians, who had invented geometry, thought little
about mathematics. For them it was a tool to reckon the passage of the days and to maintain plots of land. The Greeks
had a very different attitude. To them, numbers and philosophy were inseparable, and they took both very seriously. Indeed, the Greeks went overboard when it came to numbers.
Literally.

Hippasus of Metapontum stood on the deck, preparing to
die. Around him stood the members of a cult, a secret brotherhood that he had betrayed. Hippasus had revealed a secret
that was deadly to the Greek way of thinking, a secret that
threatened to undermine the entire philosophy that the brotherhood had struggled to build. For revealing that secret, the
great Pythagoras himself sentenced Hippasus to death by
drowning. To protect their number-philosophy, the cult
would kill. Yet as deadly as the secret that Hippasus revealed
was, it was small compared to the dangers of zero.

The leader of the cult was Pythagoras, an ancient radical.
According to most accounts, he was born in the sixth century BC on Samos, a Greek island off the coast of Turkey
famed for a temple to Hera and for really good wine. Even by
the standards of the superstitious ancient Greeks, Pythagoras's beliefs were eccentric. He was firmly convinced that he
was the reincarnated soul of Euphorbus, a Trojan hero. This
helped convince Pythagoras that all souls—including those of
animals—transmigrated to other bodies after death. Because

---

* The Greek word for *ratio* was λογος (logos), which is also the term for
word. This translation is even more rational than the traditional one.

of this, he was a strict vegetarian. Beans, however, were taboo, as they generate flatulence and are like the genitalia.

Pythagoras may have been an ancient New Age thinker, but he was a powerful orator, a renowned scholar, and a charismatic teacher. He was said to have written the constitution for Greeks living in Italy. Students flocked to him, and he soon acquired a retinue of followers who wanted to learn from the master.

The Pythagoreans lived according to the dicta of their leader. Among other things they believed that it is best to make love to women in the winter, but not in the summer; that all disease is caused by indigestion; that one should eat raw food and drink only water; and that one must avoid wearing wool. But at the center of their philosophy was the most important tenet of the Pythagoreans: all is number.

The Greeks had inherited their numbers from the geometric Egyptians. As a result, in Greek mathematics there was no significant distinction between shapes and numbers. To the Greek philosopher-mathematicians they were pretty much the same thing. (Even today, we have *square* numbers and *triangular* numbers thanks to their influence [Figure 5].) In those days, proving a mathematical theorem was often as simple as drawing an elegant picture; the tools of ancient Greek mathematics weren't pencil and paper—they were a straightedge and compasses. And to Pythagoras the connection between shapes and numbers was deep and mystical. Every number-shape had a hidden meaning, and the most beautiful number-shapes were sacred.

The mystical symbol of the Pythagorean cult was, naturally, a number-shape: the pentagram, a five-pointed star. This simple figure is a glimpse at the infinite. Nestled within the lines of the star is a pentagon. Connecting the corners of that pentagon with lines generates a small, upside-down, five-pointed star, which is exactly the same as the original star in its proportions. This star, in turn, contains an even smaller pentagon, which contains a tinier star with its tiny pentagon,

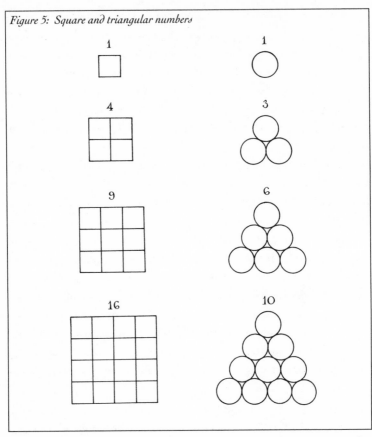

Figure 5: Square and triangular numbers

and so forth (Figure 6). As interesting as this was, to the Pythagoreans the most important property of the pentagram was not in this self-replication but was hidden within the lines of the star. They contained a number-shape that was the ultimate symbol of the Pythagorean view of the universe: the golden ratio.

The importance of the golden ratio comes from a Pythagorean discovery that is now barely remembered. In modern schools, children learn of Pythagoras

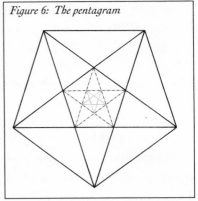

Figure 6: The pentagram

for his famed theorem: the square of the hypotenuse of a right triangle is equal to the sum of the squares of the other two sides. However, this was in fact ancient news. It was known more than 1,000 years before Pythagoras's time. In ancient Greece, Pythagoras was remembered for a different invention: the musical scale.

One day, according to legend, Pythagoras was toying with a monochord, a box with a string on it (Figure 7). By moving a sliding bridge up and down the monochord, Pythagoras changed the notes that the device played. He quickly discovered that strings have a peculiar, yet predictable, behavior. When you pluck the string without the bridge, you get a clear note, the tone known as the *fundamental*. Putting the bridge on the monochord so it touches the string changes the notes that are played. When you place the bridge exactly in the middle of the monochord, touching the center of the string, each half of the string plays the same note: a tone exactly one octave higher than the string's fundamental. Shifting the bridge slightly might divide the string so that one side has three-fifths of the string and the other has two-fifths; in this case Pythagoras noticed that plucking the string segments creates two notes that form a *perfect fifth*, which is said to be the most powerful and evocative musical relationship. Different ratios gave different tones that could soothe or disturb. (The discordant tritone, for instance, was dubbed the "devil in music" and was rejected by medieval musicians.) Oddly enough, when Pythagoras put the bridge at a place that did not divide the string into a simple ratio, the plucked notes did not meld well. The sound was usually dissonant and sometimes even worse. Often the tone wobbled like a drunkard up and down the scale.

To Pythagoras, playing music was a mathematical act. Like squares and triangles, lines were number-shapes, so dividing a string into two parts was the same as taking a ratio of two numbers. The harmony of the monochord was the harmony of mathematics—and the harmony of the universe. Pythagoras concluded that ratios govern not only music but

Figure 7: The mystical monochord

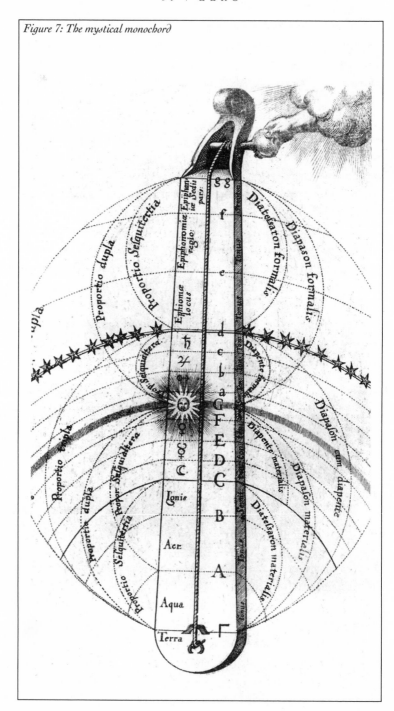

also all other types of beauty. To the Pythagoreans, ratios and proportions controlled musical beauty, physical beauty, and mathematical beauty. Understanding nature was as simple as understanding the mathematics of proportions.

This philosophy—the interchangeability of music, math, and nature—led to the earliest Pythagorean model of the planets. Pythagoras argued that the earth sat at the center of the universe, and the sun, moon, planets, and stars revolved around the earth, each pinned inside a sphere (Figure 8). The ratios of the sizes of the spheres were nice and orderly, and as the spheres moved, they made music. The outermost planets, Jupiter and Saturn, moved the fastest and made the highest-pitched notes. The innermost ones, like the moon, made

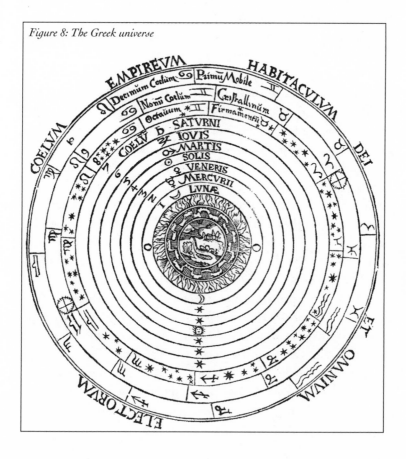

Figure 8: The Greek universe

lower notes. Taken all together, the moving planets made a "harmony of the spheres," and the heavens are a beautiful mathematical orchestra. This is what Pythagoras meant when he insisted, "All is number."

Because ratios were the keys to understanding nature, the Pythagoreans and later Greek mathematicians spent much of their energy investigating their properties. Eventually, they categorized proportions into 10 different classes, with names like the *harmonic mean*. One of these means yielded the most "beautiful" number in the world: the golden ratio.

Achieving this blissful mean is a matter of dividing a line in a special way: divide it in two so that the ratio of the small part to the large part is the same as the ratio of the large part to the whole (see appendix B). In words, it doesn't seem particularly special, but figures imbued with this golden ratio seem to be the most beautiful objects. Even today, artists and architects intuitively know that objects that have this ratio of length to width are the most aesthetically pleasing, and the ratio governs the proportions of many works of art and architecture. Some historians and mathematicians argue that the Parthenon, the majestic Athenian temple, was built so that the golden ratio is visible in every aspect of its construction. Even nature seems to have the golden ratio in its design plans. Compare the ratios of the size of any two succeeding chambers of the nautilus, or take the ratio of clockwise grooves to counterclockwise grooves in the pineapple, and you will see that these ratios approach the golden ratio (Figure 9).

The pentagram became the most sacred symbol of the Pythagorean brotherhood because the lines of the star are divided in this special way—the pentagram is chock-full of the golden ratio—and for Pythagoras, the golden ratio was the king of numbers. The golden ratio was favored by artists and nature alike and seemed to prove the Pythagorean assertion that music, beauty, architecture, nature, and the very con-

Figure 9: The Parthenon, the chambered nautilus, and the golden ratio

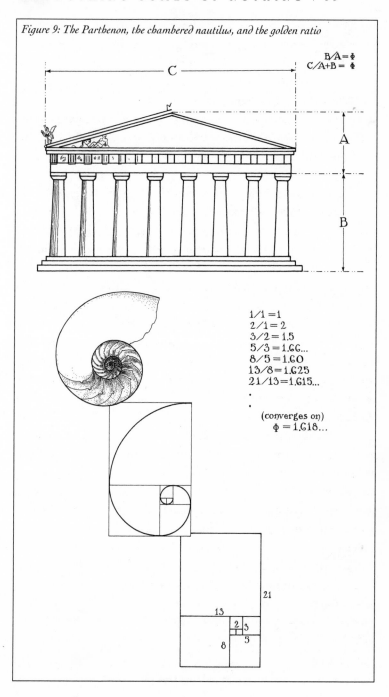

B/A = Φ
C/A+B = Φ

C

A

B

1/1 = 1
2/1 = 2
3/2 = 1.5
5/3 = 1.66...
8/5 = 1.60
13/8 = 1.625
21/13 = 1.615...

·

·

(converges on)
Φ = 1.618...

21

13

2 3

8 5

struction of the cosmos were all intertwined and inseparable. To the Pythagorean mind, ratios controlled the universe, and what was true for the Pythagoreans soon became true for the entire West. The supernatural link between aesthetics, ratios, and the universe became one of the central and long-lasting tenets of Western civilization. As late as Shakespeare's time, scientists talked about the revolution of orbs of different proportions and discussed the heavenly music that reverberated throughout the cosmos.

Zero had no place within the Pythagorean framework. The equivalence of numbers and shapes made the ancient Greeks the masters of geometry, yet it had a serious drawback. It precluded anyone from treating zero as a number. What shape, after all, could zero be?

It is easy to visualize a square with width two and height two, but what is a square with width zero and height zero? It's hard to imagine something with no width and no height — with no substance at all — being a square. This meant that multiplication by zero didn't make any sense either. Multiplying two numbers was equivalent to taking an area of a rectangle, but what could the area of a rectangle with zero height or zero width be?

Nowadays the great unsolved problems in mathematics are stated in terms of conjectures that mathematicians are unable to prove. In ancient Greece, however, number-shapes inspired a different way of thinking. The famous unsolved problems of the day were geometric: With only a straightedge and compasses, could you make a square equal in area to a given circle? Could you use those tools to trisect an angle?* Geometric constructions and shapes were the same thing. Zero was a number that didn't seem to make any geometric

---

* The early Babylonians were apparently unaware of the difficulty in trisecting an angle. In the *Epic of Gilgamesh*, the narrator states that Gilgamesh was two-thirds god and one-third man. This is as impossible as trisecting an angle with a straightedge and compasses — unless gods and mortals are allowed to have an infinite amount of sex.

sense, so to include it, the Greeks would have had to revamp their entire way of doing mathematics. They chose not to.

Even if zero were a number in the Greek sense, the act of taking a ratio with zero in it would seem to defy nature. No longer would a proportion be a relationship between two objects. The ratio of zero to anything—zero divided by a number—is always zero; the other number is completely consumed by the zero. And the ratio of anything to zero—a number divided by zero—can destroy logic. Zero would punch a hole in the neat Pythagorean order of the universe, and for that reason it could not be tolerated.

Indeed, the Pythagoreans had tried to squelch another troublesome mathematical concept—the irrational. This concept was the first challenge to the Pythagorean point of view, and the brotherhood tried to keep it secret. When the secret leaked out, the cult turned to violence.

The concept of the irrational was hidden like a time bomb inside Greek mathematics. Thanks to the number-shape duality, to the Greeks counting was tantamount to measuring a line. Thus, a ratio of two numbers was nothing more than the comparison of two lines of different lengths. However, to make any sort of measurement, you need a standard, a common yardstick, to compare to the size of the lines. As an example, imagine a line exactly a foot long. Make a mark, say, five and a half inches from one end, dividing the foot into two unequal parts. The Greeks would figure out the ratio by dividing the line into tiny pieces, using, for example, a standard or yardstick half an inch long. One line segment contains 11 of those pieces; the other contains 13. The ratio of the two segments, then, is 11 to 13.

For everything in the universe to be governed by ratios, as the Pythagoreans hoped, everything that made sense in the universe had to be related to a nice, neat proportion. It literally had to be *rational*. More precisely, these ratios had to be written in the form *a/b*, where *a* and *b* were nice, neat counting numbers like 1, 2, or 47. (Mathematicians are careful to

note that *b* is not allowed to be zero, for that would be tantamount to division by zero, which we know to be disastrous.)
Needless to say, the universe is not really that orderly. Some
numbers cannot be expressed as a simple ratio of *a/b*. These
irrational numbers were an unavoidable consequence of
Greek mathematics.

The square is one of the simplest figures of geometry, and
it was duly revered by the Pythagoreans. (It had four sides,
corresponding to the four elements; it symbolized the perfec-
tion of numbers.) But the irrational is nestled within the simplicity of the square. If you draw the diagonal—a line from
one corner to the opposite corner—the irrational appears. As a
concrete example, imagine a square whose sides are one foot
long. Draw the diagonal. Ratio-obsessed people like the Greeks
naturally looked at the side of the square and the diagonal
and asked themselves: what is the ratio of the two lines?

The first step, again, is to create a common yardstick, perhaps a tiny ruler half an inch long. The next step is to use that
yardstick to divide each of the two lines into equal segments.
With a half-inch yardstick we can divide the foot-long side of
the square into 24 segments, each half an inch long. What
happens when we measure the diagonal? Using the same
yardstick, we see that the diagonal gets . . . well, almost 34
segments, but it doesn't come out quite evenly. The 34th segment is a wee bit too short; the half-inch ruler juts out a little
beyond the corner of the square. We can do better. Let's divide the line into even smaller segments, using, say, a ruler
one-sixth of an inch long. The side of the square is partitioned
into 72 segments, while the diagonal comes out to more than
101 but fewer than 102 segments. Again, the measurement is
not quite perfect. What happens when we try *really* small segments, measuring in bits a millionth of an inch each? The side
of the square gets 12 million bits, and the diagonal gets a tad
less than 16,970,563 bits. Once again, our ruler doesn't fit
both lines exactly. No matter what ruler we choose, our measurement never seems to come out right.

In fact, no matter how tiny you make the bits, it is impossible to choose a common yardstick that will measure both the side and the diagonal perfectly: the diagonal is *incommensurable* with the side. However, without a common yardstick, it is impossible to express the two lines in a ratio. For a square of size one, this means that we cannot choose counting numbers *a* and *b* such that the diagonal of the square can be expressed as *a/b*. In other words, the diagonal of that square is *irrational*—and nowadays we recognize that number as the square root of two.

This meant trouble for the Pythagorean doctrine. How could nature be governed by ratios and proportions when something as simple as a square can confound the language of ratios? This idea was hard for the Pythagoreans to believe, but it was incontrovertible—a consequence of the mathematical laws that they held so dear. One of the first mathematical proofs in history was about the incommensurability/irrationality of the square's diagonal.

Irrationality was dangerous to Pythagoras, as it threatened the basis of his ratio-universe. To add insult to injury, the Pythagoreans soon discovered that the golden ratio, the ultimate Pythagorean symbol of beauty and rationality, was an irrational number. To keep these horrible numbers from ruining the Pythagorean doctrine, the irrationals were kept secret. Everyone in the Pythagorean brotherhood was already tight-lipped—nobody was allowed even to take written notes—and the incommensurability of the square root of two became the deepest, darkest secret of the Pythagorean order.

However, irrational numbers, unlike zero, could not easily be ignored by the Greeks. The irrationals occurred and reoccurred in all sorts of geometrical constructions. It would be hard to keep the secret of the irrational hidden from a people so obsessed with geometry and ratios. One day someone was going to let the secret out. This someone was Hippasus of Metapontum, a mathematician and member of the Pythagorean brotherhood. The secret of the irrationals would cause him great misfortune.

The legends are very hazy and contain contradictory stories about the betrayal and ultimate fate of Hippasus. Mathematicians to this day tell of the hapless man who revealed the secret of the irrational to the world. Some say that the Pythagoreans tossed Hippasus overboard, drowning him, a just punishment for ruining a beautiful theory with harsh facts. Ancient sources talk about his perishing at sea for his impiety, or alternatively, say that the brotherhood banished him and constructed a tomb for him, expelling him from the world of human beings. But whatever Hippasus's true fate was, there is little doubt that he was reviled by his brothers. The secret he revealed shook the very foundations of the Pythagorean doctrine, but by considering the irrational an anomaly, the Pythagoreans could keep the irrationals from contaminating their view of the universe. Indeed, over time the Greeks reluctantly admitted the irrationals to the realm of numbers. The irrationals didn't kill Pythagoras. Beans did.

Just as the legends of Hippasus's murder are hazy, so too are the legends of Pythagoras's end. Nevertheless, they all imply that the master died in a bizarre way. Some say that Pythagoras starved himself, but the most common versions all say that beans were his undoing. One day, according to a version of the legend, his house was set ablaze by his enemies (who were angry at not being considered worthy to be admitted into Pythagoras's presence), and the brothers in the house scattered in all directions, running for their lives. The mob slaughtered Pythagorean after Pythagorean. The brotherhood was being destroyed. Pythagoras himself fled for his life, and he might have gotten away had he not run smack into a bean field. There he stopped. He declared that he would rather be killed than cross the field of beans. His pursuers were more than happy to oblige. They cut his throat.

Though the brotherhood was scattered and the leader was dead, the essence of the Pythagorean teachings lived on. It was soon to become the basis of the most influential philosophy in Western history—the Aristotelian doctrine that would live for two millennia. Zero would clash with this doctrine, and unlike the irrational, zero could be ignored. The number-shape duality in Greek numbers made it easy; after all, zero didn't have a shape and could thus not be a number.

Yet it was not the Greek number system that prevented zero's acceptance — nor was it lack of knowledge. The Greeks had learned about zero because of their obsession with the night sky. Like most ancient peoples, the Greeks were stargazers, and the Babylonians were the first masters of astronomy: they had learned how to predict eclipses. Thales, the first Greek astronomer, learned how to do this from the Babylonians, or perhaps through the Egyptians. In 585 BC he was said to have predicted a solar eclipse.

With Babylonian astronomy came Babylonian numbers. For astronomical purposes the Greeks adopted a sexagesimal number system and even divided hours into 60 minutes, and minutes into 60 seconds. Around 500 BC the placeholder zero began to appear in Babylonian writings; it naturally spread to the Greek astronomical community. During the peak of ancient astronomy, Greek astronomical tables regularly employed zero; its symbol was the lowercase omicron, *o*, which looks very much like our modern-day zero, though it's probably a coincidence. (Perhaps the use of omicron came from the first letter of the Greek word for nothing, *ouden*.) The Greeks didn't like zero at all and used it as infrequently as possible. After doing their calculations with Babylonian notation, Greek astronomers usually converted the numbers back into clunky Greek-style numerals — without zero. Zero never worked its way into ancient Western numbers, so it is unlikely that the omicron is the mother of our 0. The Greeks saw the usefulness of zero in their calculations, yet they still rejected it.

So it was not ignorance that led the Greeks to reject zero, nor was it the restrictive Greek number-shape system. It was philosophy. Zero conflicted with the fundamental philosophical beliefs of the West, for contained within zero are two ideas that were poisonous to Western doctrine. Indeed, these concepts would eventually destroy Aristotelian philosophy after its long reign. These dangerous ideas are the void and the infinite.

The Infinite, the Void, and the West

So, naturalists observe, a flea

Hath smaller fleas that on him prey,

And these have smaller yet to bite 'em,

And so proceed ad infinitum. . . .

— JONATHAN SWIFT,
"ON POETRY: A RHAPSODY"

The infinite and the void had powers that frightened the Greeks. The infinite threatened to make all motion impossible, while the void threatened to smash the nutshell universe into a thousand flinders. By rejecting zero, the Greek philosophers gave their view of the universe the durability to survive for two millennia.

Pythagoras's doctrine became the centerpiece of Western philosophy: all the universe was governed by ratios and shapes; the planets moved in heavenly spheres that made music as they turned. But what lay beyond these spheres? Were there more and more spheres, each larger than its neighbor? Or was the outermost sphere the end of the universe? Aristotle and later philosophers would insist that there could not be an infinite number of nested spheres. With the adoption of this philosophy, the West had no room for infinity or the infinite. They rejected it outright. For the infinite had already begun to gnaw at the roots of Western thought, thanks to Zeno of Elea, a philosopher reckoned by his contemporaries to be the most annoying man in the West.

Zeno was born around 490 BC, at the beginning of the Persian wars — a great conflict between East and West. Greece would defeat the Persians; Greek philosophy would never quite defeat Zeno — for Zeno had a paradox, a logical puzzle that seemed intractable to the reasoning of Greek philosophers. It was the most troubling argument in Greece: Zeno had proved the impossible.

According to Zeno, nothing in the universe could move. Of course, this is a silly statement; anyone can refute it by walking across the room. Though everybody knew that Zeno's statement was false, nobody could find a flaw in Zeno's argument. He had come up with a paradox. Zeno's logical puzzle baffled Greek philosophers—as well as the philosophers who came after them. Zeno's riddles plagued mathematicians for nearly two thousand years.

In his most famous puzzle, "The Achilles," Zeno proves that swift Achilles can never catch up with a lumbering tortoise that has a head start. To make things more concrete, let's put some numbers on the problem. Imagine that Achilles runs at a foot a second, while the tortoise runs at half that speed. Imagine, too, that the tortoise starts off a foot ahead of Achilles.

Achilles speeds ahead, and in a mere second he has caught up to where the tortoise was. But by the time he reaches that point, the tortoise, which is also running, has moved ahead by half a foot. No matter. Achilles is faster, so in half a second, he makes up the half foot. But again, the tortoise has moved ahead, this time by a quarter foot. In a flash—a quarter second—Achilles has made up the distance. But the tortoise lumbers ahead in that time by an eighth of a foot. Achilles runs and runs, but the tortoise scoots ahead each time; no matter how close Achilles gets to the tortoise, by the time he reaches the point where the tortoise was, the tortoise has moved. An eighth of a foot . . . a sixteenth of a foot . . . a thirty-second of a foot . . . smaller and smaller distances, but Achilles never catches up. The tortoise is always ahead (Figure 10).

Everybody knows that, in the real world, Achilles would quickly run past the tortoise, but Zeno's argument seemed to prove that Achilles could never catch up. The philosophers of his day were unable to refute the paradox. Even though they knew that the conclusion was wrong, they could never find a mistake in Zeno's mathematical proof. The philosophers' main weapon was logic, but logical deduction seemed useless

Figure 10: Achilles and the tortoise

against Zeno's argument. Each step along the way seemed airtight, and if all the steps are correct, how could the conclusion be wrong?

The Greeks were stumped by the problem, but they did find the source of the trouble: infinity. It is the infinite that lies at the heart of Zeno's paradox: Zeno had taken continuous motion and divided it into an infinite number of tiny steps. Because there are an infinite number of steps, the Greeks assumed that the race would go on forever and ever, even though the steps get smaller and smaller. The race would never finish in finite time—or so they thought. The ancients didn't have the equipment to deal with the infinite, but modern mathematicians have learned to handle it. The infinite must be approached very carefully, but it can be mastered, with the help of zero. Armed with 2,400 years' worth of extra mathematics, it is not hard for us to go back and find Zeno's Achilles' heel.

The Greeks did not have zero, but we do, and it is the key to solving Zeno's puzzle. It is sometimes possible to add infinite terms together to get a finite result—but to do so, the terms being added together must approach zero.* This is the case with Achilles and the tortoise. When you add up the distance that Achilles runs, you start with the number 1, then add 1/2, then add 1/4, then add 1/8, and so on, with the terms getting smaller and smaller, getting closer and closer to zero; each term is like a step along a journey where the destination is zero. However, since the Greeks rejected the number zero, they couldn't understand that this journey could ever have an end. To them, the numbers 1, 1/2, 1/4, 1/8, 1/16, and so forth aren't approaching anything; the destination doesn't exist. Instead, the Greeks just saw the terms as simply getting smaller and smaller, meandering outside the realm of numbers.

Modern mathematicians know that the terms have a *limit;* the numbers 1, 1/2, 1/4, 1/8, 1/16, and so forth are approaching zero as their limit. The journey has a destination. Once the journey has a destination, it is easy to ask how far away that destination is and how long it will take to get there. It is not that difficult to sum up the distances that Achilles runs: $1 + 1/2 + 1/4 + 1/8 + 1/16 + \ldots + 1/2^n + \ldots$. In the same way that the steps that Achilles takes get smaller and smaller, and closer and closer to zero, the sum of those steps gets closer and closer to 2. How do we know this? Well, let's start off with 2, and subtract the terms of the sum, one by one. We begin with $2 - 1$, which is, of course, 1. Next, we subtract 1/2, leaving 1/2. Then remove the next term: subtract 1/4, leaving

---

* This is a necessary, but not sufficient, condition. If the terms go to zero too slowly, then the sum of the terms doesn't *converge* to a finite number.

1/4 behind. Subtracting 1/8 leaves 1/8 behind. We're back to our familiar sequence. We already know that 1, 1/2, 1/4, 1/8, and so forth has a limit of zero; thus, as we subtract the terms from 2, we have nothing left. The limit of the sum $1 + 1/2 + 1/4 + 1/8 + 1/16 + \ldots$ is 2 (Figure 11). Achilles runs 2 feet in catching up to the tortoise, even though he takes an infinite number of steps to do it. Better yet, look at the time it takes Achilles to overtake the tortoise: $1 + 1/2 + 1/4 + 1/8 + 1/16 + \ldots -2$ seconds. Not only does Achilles take an infinite number of steps to run a finite distance, but he takes only 2 seconds to do it.

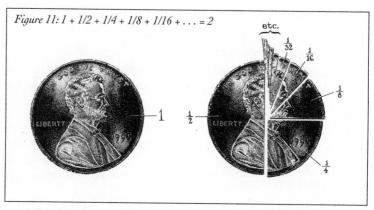

Figure 11: $1 + 1/2 + 1/4 + 1/8 + 1/16 + \ldots = 2$

The Greeks couldn't do this neat little mathematical trick. They didn't have the concept of a limit because they didn't believe in zero. The terms in the infinite series didn't have a limit or a destination; they seemed to get smaller and smaller without any particular end in sight. As a result, the Greeks couldn't handle the infinite. They pondered the concept of the void but rejected zero as a number, and they toyed with the concept of the infinite but refused to allow infinity—numbers that are infinitely small and infinitely large—anywhere near the realm of numbers. This is the biggest failure in Greek mathematics, and it is the only thing that kept them from discovering calculus.

Infinity, zero, and the concept of limits are all tied together in a bundle. Greek philosophers were unable to untie that bundle; therefore, they were ill-equipped to solve Zeno's puzzle. Yet Zeno's paradox was so powerful that the Greeks tried over and over to explain away his infinities. They were doomed to failure, unarmed with the proper concepts.

Zeno himself didn't have a proper solution to the paradox, nor did he seek one. The paradox suited his philosophy perfectly. He was a member of the Eleatic school of thought, whose founder, Parmenides, held that the underlying nature of the universe was changeless and immobile. Zeno's puzzles appear to have been in support of Parmenides' argument; in showing that change and motion were paradoxical, he hoped to convince people that everything is one—and changeless. Zeno really did believe that motion was impossible, and his paradox was this theory's chief support.

There were other schools of thought. The atomists, for example, believed that the universe is made up of little particles called atoms, which are indivisible and eternal. Motion, according to the atomists, was the movement of these little particles. Of course, for these atoms to move, there has to be empty space for them to move into. After all, these little atoms had to move around somehow; if there were no such thing as a vacuum, the atoms would be constantly pressed against one another. Everything would be stuck in one position for eternity, unable to move. Thus, the atomic theory required that the universe be filled with emptiness—an infinite void. The atomists embraced the concept of the infinite vacuum—infinity and zero wrapped into one. This was a shocking conclusion, but the indivisible kernels of matter in atomic theory got around the problem of Zeno's paradoxes. Because atoms are indivisible, there is a point beyond which things could not be divided. Zeno's hair-splitting doesn't go on ad infinitum. After a number of strides, Achilles would be taking tiny steps that can't get any smaller; eventually he would have to hurdle an atom that the tortoise doesn't. Achilles would finally catch up to the elusive turtle.

Another philosophy vied with the atomic theory, and instead of posing such bizarre concepts as the infinite vacuum, it turned the universe into a cozy nutshell. There was no infinity, no void—just beautiful spheres that surrounded the earth, which was naturally placed at the very center of the universe. This was the Aristotelian system, which was later refined by the Alexandrian astronomer Ptolemy. It became the dominant philosophy in the Western world. And by rejecting zero and infinity, Aristotle explained away Zeno's paradoxes.

Aristotle simply declared that mathematicians "do not need the infinite, or use it." Though "potential" infinities could exist in the minds of mathematicians—like the concept of dividing lines into infinite pieces—nobody could actually do it, so the infinite doesn't exist in reality. Achilles runs smoothly past the tortoise because the infinite points are simply a figment of Zeno's imagination, rather than a real-world construct. Aristotle just wished infinity away by stating that it is simply a construct of the human mind.

From that concept comes a startling revelation. Based upon the Pythagorean universe, the Aristotelian cosmos (and its later refinement by the astronomer Ptolemy) had the planets moving in crystalline orbs. However, since there is no infinity, there can't be an endless number of spheres; there must be a last one. This outermost sphere was a midnight blue globe encrusted with tiny glowing points of light—the stars. There was no such thing as "beyond" the outermost sphere; the universe ended abruptly with that outermost layer. The universe was contained in a nutshell, ensconced comfortably within the sphere of fixed stars; the cosmos was finite in extent, and entirely filled with matter. There was no infinite; there was no void. There was no infinity; there was no zero.

This line of reasoning had another consequence—and this is why Aristotle's philosophy endured for so many years. His system proved the existence of God.

The heavenly spheres are slowly spinning in their places, making a music that suffuses the cosmos. But something must be causing that motion. The stationary earth cannot be the source of that motive power, so the innermost sphere must be moved by the next sphere out. That sphere, in turn, is moved by its larger neighbor, and on and on. However, there is no infinity; there are a finite number of spheres, and a finite number of things that are moving each other. Something must be the ultimate cause of motion. Something must be moving the sphere of fixed stars. This is the prime mover: God. When Christianity swept through the West, it became closely tied to the Aristotelian view of the universe and the proof of God's existence. Atomism became associated with atheism. Questioning the Aristotelian doctrine was tantamount to questioning God's existence.

Aristotle's system was extremely successful. His most famous student, Alexander the Great, spread the doctrine as far east as India before Alexander's untimely death in 323 BC. The Aristotelian system would outlast Alexander's empire; it would survive until Elizabethan times, the sixteenth century. With this long-standing acceptance of Aristotle came a rejection of the infinite—and the void, for Aristotle's denial of the infinite required a denial of the void, because the void implies the existence of the infinite. After all, there were only two logical possibilities for the nature of the void, and both implied that the infinite exists. First, there could be an infinite amount of void—thus infinity exists. Second, there could be a finite amount of void, but since void is simply the lack of matter, there must be an infinite amount of matter to make sure that there is only a finite amount of void—thus infinity exists. In both cases the existence of the void implies the existence of the infinite. Void/zero destroys Aristotle's neat argument, his refutation of Zeno, and his proof of God. So as Aristotle's arguments were accepted, the Greeks were forced to reject zero, void, the infinite, and infinity.

There was a problem, though. It is not so easy to reject both infinity and zero. Look back through time. Events have happened throughout history, but if there is no such thing as infinity, there cannot be an infinite number of events. Thus, there must be a first event: creation. But what existed before creation? Void? That was unacceptable to Aristotle. Conversely, if there was not a first event, then the universe must have always existed—and will always exist in the future.

You've got to have either infinity or zero; a universe without both of them makes no sense.

Aristotle hated the idea of the void so much that he chose the eternal, infinite universe over one that had a vacuum in it; he said that the eternity of time was a "potential" infinity like Zeno's infinite subdivisions. (It was a stretch, but many scholars bought the argument; some even chose the creation story as further evidence for God. Medieval philosophers and theologians were doomed to battle over this puzzle for several hundred years.)

The Aristotelian view of physics, as wrong as it was, was so influential that for more than a millennium it eclipsed all opposing views, including more realistic ones. Science would never progress until the world discarded Aristotle's physics — along with Aristotle's rejection of Zeno's infinities.

For all his wit, Zeno got himself into serious trouble. Around 435 BC, he conspired to overthrow the tyrant of Elea, Nearchus. He was smuggling arms to support the cause. Unfortunately for Zeno, Nearchus found out about the plot, and Zeno was arrested. Hoping to discover who the coconspirators were, Nearchus had Zeno tortured. Soon Zeno begged the torturers to stop and promised that he would name his colleagues. When Nearchus drew near, Zeno insisted that the tyrant come closer, since it was best to keep the names a secret. Nearchus leaned over, tilting his head toward Zeno. All of a sudden Zeno sank his teeth into Nearchus's ear. Nearchus screamed, but Zeno refused to let go. The torturers could only force Zeno to let go by stabbing him to death. Thus died the master of the infinite.

Eventually, one ancient Greek surpassed Zeno in matters of the infinite: Archimedes, the eccentric mathematician of Syracuse. He was the only thinker of his day to glimpse the infinite.

Syracuse was the richest city on the island of Sicily, and Archimedes was its most famous resident. Little is known about his youth, but it seems that Archimedes was born around 287 BC in Samos, Pythagoras's birthplace. He then immigrated to Syracuse, where he solved engineering problems for the king. It was the king of Syracuse who asked Archimedes to determine whether his crown was pure gold or had been mixed with lead, a task beyond the abilities of all the scientists at the time. However, when Archimedes settled into a tub of water, he noticed that the water flowed over the sides, and he suddenly realized that he could measure the density of the crown, and thus its purity, by submerging it in a tub of water and measuring how much water it displaced. Elated by the insight, Archimedes leapt out of the tub and ran through the streets of Syracuse shouting, "Eureka! Eureka!" Of course Archimedes forgot that he was stark naked.

Archimedes' talents were useful to the Syracusan military as well. In the third century BC the era of Greek hegemony was over. Alexander's empire had collapsed into bickering states, and a new power was flexing its muscles in the West: Rome. And Rome had set its sights on Syracuse. According to legend, Archimedes armed the Syracusans with miraculous weapons to defend the city from the Romans: stone throwers; huge cranes that grabbed the Roman warships, lifted them up, and dumped them, bow first, into the water; and mirrors made of such quality that they set Roman ships afire at a great distance by reflecting sunlight. The Roman soldiers were so afraid of these war engines that if they saw so much as a bit of rope or wood sticking over the wall they would flee for fear that Archimedes was aiming a weapon at them.

Archimedes first glimpsed the infinite in the polish of his war mirrors. For centuries the Greeks had been fascinated with conic sections. Take a cone and cut it up; you get circles, ellipses, parabolas, and hyperbolas, depending on how you slice it. The parabola has a special property: it takes the rays of light from the sun, or any distant source, and focuses them to a point, concentrating all the light's energy on a very small area. Any mirror that could set ships afire must be in the shape of a parabola. Archimedes studied the properties of the parabola, and it is here that he first started toying with the infinite.

To understand the parabola, Archimedes had to learn how to measure it; for instance, nobody knew how to determine the area of a section of a parabola. Triangles and circles were easy to measure, but slightly more irregular curves like the parabola were beyond the ken of the Greek mathematicians of the day. However, Archimedes figured out a way to measure the parabola's area by resorting to the infinite. The first step was to inscribe a triangle inside the parabola. In the two little gaps left, Archimedes inscribed more triangles. This left four gaps, which were filled with more triangles, and so on (Figure 12). It's like Achilles and the tortoise—an infinite series of steps, each getting smaller and smaller. The areas of the little triangles quickly approach zero. After a long, involved set of calculations, Archimedes summed the areas of the infinite triangles and divined the area of the parabola. However, any mathematician of the day would have scoffed at this line of reasoning; Archimedes used the tools of the infinite, which were so expressly disallowed by his mathematical colleagues. To satisfy them, Archimedes also included a proof, based upon the accepted mathematics of the time, that relied upon the so-called axiom of Archimedes, although Archimedes himself mentioned that earlier mathematicians deserved the credit. As you may recall, this axiom says that any number added to

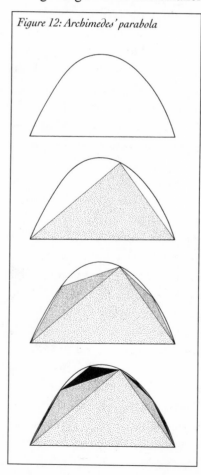

Figure 12: Archimedes' parabola

itself over and over again can exceed any other number. Zero, clearly, was not included.

Archimedes' proof by triangles was as close as you could come to the idea of limits—and calculus—without actually discovering them. In later works Archimedes figured out the volumes of parabolas and circles, rotated around a line, which any math student knows are early homework problems in a calculus course. But the axiom of Archimedes rejected zero, which is the bridge between the realms of the finite and the infinite, a bridge that is absolutely necessary for calculus and higher mathematics.

Even the brilliant Archimedes occasionally scorned the infinite along with his contemporaries. He believed in the Aristotelian universe; the universe was contained within a giant sphere. On a whim he decided to calculate how many grains of sand could fit in the (spherical) universe. In his "Sand Reckoner," Archimedes first calculated how many grains of sand would fit across a poppy seed, and then how many poppy seeds would span a finger's breadth. From finger breadths to the length of a stadium (the standard Greek unit of long distance) and on to the size of the universe, Archimedes figured out that 10^{51} grains of sand would fill the entire universe, packed full even unto the outermost sphere of fixed stars. (10^{51} is a really, really big number. Take 10^{51} molecules of water, for instance. It would take every man, woman, and child now on Earth, each drinking a ton of water a second, over 150,000 years to drink it all.) This number was so large that the Greek system of numeration was unable to handle it; Archimedes had to invent a whole new method of denoting really huge numbers.

The myriad was the biggest grouping in the Greek number system, and by counting myriads the Greeks could denote numbers up to a myriad myriads (100,000,000) and a little further. But Archimedes got beyond this limitation by hitting the reset button. He simply started over at a myriad myriads, setting 100,000,000 equal to one, and counting again, calling these new numbers, numbers of the *second order*. (Archimedes did not set 100,000,001 equal to one and 100,000,000 equal to zero, which is what a modern mathematician would do. It had never occurred to Archimedes that starting over with zero would make more sense.) The numbers of the second order went from one myriad myriads to a myriad myriad myriad myriads. The numbers of the third order went to one myriad myriad myriad myriad myriad myriads (1,000,000,000,000,000,000,000,000), and so forth, until he reached the numbers of the myriad myriadth order, which he called the numbers of the first period. It was a cumbersome way of doing business, but it got the job done and went far beyond what Archimedes needed to solve his thought experiment. Yet as big as those numbers were, they were finite — and were more than enough to fill the universe to overflowing with sand. The infinite was not needed in the Greek universe.

Perhaps, given more time, Archimedes might have begun to see the lure of the infinite and of zero. But the sand reckoner was destined to meet his fate while reckoning in the sand. The Romans were too powerful for the Syracusans. Taking advantage of a poorly manned watchtower and an easy-to-climb wall, the Romans managed to get some soldiers inside the city. As soon as they realized that Romans were inside the walls of the city, the Syracusans, wild with fear, could not mount a defense. The Romans poured through the city, but Archimedes was deaf to the panic around him. He sat on the ground, drawing circles in the sand, trying to prove a theorem. A Roman soldier saw the bedraggled 75-year-old and demanded that Archimedes follow him. Archimedes refused, since his mathematical proof was not yet finished. The enraged soldier cut him down. Thus died the greatest mind in the ancient world, slaughtered needlessly by the Romans.

Killing Archimedes was one of the biggest Roman contributions to mathematics. The Roman era lasted for about seven centuries. In all that time there were no significant mathematical developments. History marched on: Christianity swept through Europe, the Roman Empire fell, the Library at Alexandria burned, and the Dark Ages began. It would be another seven centuries before zero reappeared in the West. In the meantime two monks created a calendar without zero, damning us to eternal confusion.

Blind Dates

It is a silly, childish discussion and only exposes the want of brains of those who maintain a contrary opinion to that we have stated.

— THE TIMES (LONDON),
DECEMBER 26, 1799

This "silly, childish discussion"—whether the new century begins on the year 00 or the year 01—appears and reappears like clockwork every hundred years. If medieval monks had only known of zero, our calendar would not be in such a muddle.

The monks can't be faulted for their ignorance. Indeed, during the Middle Ages the only Westerners who studied math were the Christian monks. They were the only learned ones left. Monks needed math for two things: prayer and money. To count money, they needed to know how to . . . well . . . count. To do this, they used an abacus or a counting board, a device similar to an abacus, where stones or other counters get moved around on a table. It was not a very demanding task, but by ancient standards it was the state of the art. To pray, monks needed to know the time and the date. As a result, timekeeping was vital to the monks' rituals. They had different prayers to be said at different hours, day in and day out. (Our word *noon* comes from the word nones, the midday prayer service of medieval clergy.) How else would the night watchman know when to roust his fellows out of their comfortable straw beds to begin the day's devotions? And if you didn't have an accurate calendar, you couldn't know when to celebrate Easter. This was a big problem.

Calculating the date of Easter was no mean feat, thanks

to a clash of calendars. The seat of the church was Rome, and Christians used the Roman solar calendar that was 365 days (and change) long. But Jesus was a Jew, and he used the Jewish lunar calendar that was only 354 days (and change) long. The big events in Jesus' life were marked with reference to the moon, while everyday life was ruled by the sun. The two calendars drifted with respect to each other, making it very difficult to predict when a holiday was due. Easter was just such a drifting holiday, so every few generations a monk was drafted to calculate the dates when Easter would fall for the next few hundred years.

Dionysius Exiguus was one of these monks. In the sixth century the pope, John I, asked him to extend the Easter tables. While translating and recalculating the tables, Dionysius did a little research on the side; he realized that he could figure out just when Jesus Christ was born. After chugging through a bit of math, he decided that the current year was the 525th year since the birth of Christ. Dionysius decided that the year of Christ's birth should, thenceforth, be the year 1 *anno Domini,* or the first year of Our Lord. (Technically, Dionysius said that Christ's birth happened on December 25 the year before, and he started his calendar on January 1 to match the Roman year.) The next year after that was 2 AD, and the next 3 AD, and so forth, replacing the two dating systems then most commonly in use.* But there was a problem. Make that two.

For one thing, Dionysius got the date of Christ's birth wrong. The sources agree that Mary and Joseph fled the wrath of King Herod, since Herod had heard a prophecy about a newborn Messiah. But Herod died in 3 BC, years be-

---

* One dating system had the year 1 based upon the founding of the city of Rome, and the other was based on the accession of the emperor Diocletian. To the Christian monk, the birth of his Savior was a more important event than the foundation of a city that had been sacked by Vandals and Goths a few times—or, for that matter, the beginning of the reign of an emperor who had an unfortunate penchant for maintaining his menagerie of exotic animals on a diet of Christians.

fore the supposed birth of Christ. Dionysius was clearly wrong; today most scholars believe that the birth of Christ was in 4 BC. Dionysius was a few years off.

In truth, this mistake was not so terrible. When choosing the first year of a calendar, it really doesn't matter *which* year is chosen, so long as everything is consistent after that. A four-year error is inconsequential if everyone agrees to make the same mistake, as, indeed, we have. But there was a more serious problem with Dionysius's calendar: zero.

There was no year zero. Normally this would be no big deal; most calendars of that day started with the year one, not the year zero. Dionysius didn't even have a choice; he didn't know about zero. He was brought up after the decline of the Roman Empire. Even during the heyday of Rome, the Romans were not exactly math whizzes. In the year 525, at the start of the Dark Ages, all Westerners clung to the clunky Roman style of numbers, and there was no zero in that counting system. To Dionysius, the first year of Our Lord was, naturally, the year I. The next year was year II, and Dionysius came to this conclusion in the year DXXV. In most circumstances this would not have caused any trouble, especially since Dionysius's calendar did not catch on immediately. In 525 there was serious trouble for the intellectuals in the Roman court. Pope John died, and in the ensuing power shift all the philosophers and mathematicians like Dionysius were kicked out of office. They were lucky to escape with their lives. (Others were not so lucky. Anicius Boethius was a powerful courtier who was among the finest medieval Western mathematicians, which makes him worth noting. At about the same time that Dionysius was kicked out of office, Boethius, too, fell from power and was imprisoned. Boethius is not remembered for his math but for his *Consolation of Philosophy,* a tract in which he comforts himself with Aristotelian-style philosophy. He was clubbed to death soon afterward.) In any case, the new calendar languished for years.

The lack of a year zero began to cause problems two centuries later. In 731 AD, about the time Dionysius's Easter tables were set to run out, Bede, a soon-to-be-venerable monk from the northern part of England, extended them again. This is probably how he came to know of Dionysius's work. When Bede wrote a history of the church in Britain, the *Ecclesiastical History of the English People*, he used the new calendar.

The book was a huge success, but it had one significant flaw. Bede started his history with the year 60 BC—60 years before Dionysius's reference year. Bede didn't want to abandon the new dating system, so he extended Dionysius's calendar backward. To Bede, also ignorant of the number zero, the year that came before 1 AD was 1 BC. There was no year zero. After all, to Bede, zero didn't exist.

At first glance this style of numbering might not seem so bad, but it guaranteed trouble. Think of the AD years as positive numbers and the BC years as negative ones. Bede's style of counting went . . . , –3, –2, –1, 1, 2, 3, ... Zero, whose proper place is between –1 and 1, is nowhere to be seen. This throws everybody off. In 1996, an article about the calendar in the *Washington Post* told people "how to think" about the millennium controversy—and casually mentioned that since Jesus was born in 4 BC, the year 1996 was the 2,000th year since his birth. That makes perfect sense: 1996 – (– 4) = 2000. But it is wrong. It was actually only 1,999 years.

Imagine a child born on January 1 in the year 4 BC. In 3 BC he turns one year old. In 2 BC he turns two years old. In 1 BC he turns three years old. In 1 AD he turns four years old. In 2 AD he turns five years old. On January 1 in 2 AD, how many years has it been since he was born? Five years, obviously. But this isn't what you get if you subtract the years: 2 – (– 4) = 6 years old. You get the wrong answer because there is no year zero.

By rights, the child should have turned four years old on January 1 in the year 0 AD, five in 1 AD, and six in 2 AD. Then all the numbers would come out right, and figuring out the child's age would be a simple matter of subtracting – 4 from 6. But it isn't so. You've got to subtract an additional year from the total to get the right answer. Hence, Jesus was not

2,000 years old in 1996; he was only 1,999. It's very confusing, and it gets worse.

Imagine a child born in the first second of the first day of the first year: January 1 in 1 AD. In the year 2, he would be one year old, in the year 3 he would be two, and so forth; in the year 99 he'd be 98 years old, and in the year 100 he'd be 99 years old. Now imagine that this child is named Century. The century is only 99 years old in the year 100, and only celebrates its hundredth birthday on January 1 in the year 101. Thus the second century begins in the year 101. Likewise, the third century begins in the year 201, and the twentieth century begins in the year 1901. This means that the twenty-first century—and the third millennium—begins in the year 2001. Not that you'd notice.

Hotels and restaurants around the world were completely booked well in advance for December 31, 1999—not so for December 31, 2000. Everybody celebrated the turn of the millennium on the wrong date. Even the Royal Greenwich Observatory, the official keeper of the world's time and arbiter of all things chronological, planned to be swamped by the revelers. While the atomic-precision clocks ticked away in the observatory on the hill, the masses down below awaited a state-sponsored *Millennium Experience,* complete with a "spectacular opening ceremony" that the organizers scheduled for—you guessed it—December 31, 1999. The exhibit's close on December 31, 2000, is just when the astronomers on top of the hill crack open their champagne bottles to celebrate the turn of the millennium. That is, of course, assuming that astronomers care about the date at all.

Astronomers can't play with time as easily as everyone else can. After all, they are watching the clockwork of the heavens—a clockwork that does not hiccup on leap years or reset itself every time humans decide to change the calendar. Thus the astronomers decided to ignore human calendars altogether. They don't measure time in years since the birth of Christ. They count days since January 1, 4713 BC, a pretty-much arbitrary date that the scholar Joseph Scaliger chose in

1583. His *Julian Date* (named after his father, Julius, rather than Julius Caesar) became the standard way to refer to astronomical events, because it avoided all the weirdness caused by calendars that were constantly under construction. (The system has since been modified slightly. Modified Julian Date is simply the Julian Date less 2,400,000 days and 12 hours, putting the zero hour at midnight on November 17, 1858. Again, a more or less arbitrary date.) Perhaps astronomers will refuse to celebrate 51542 Modified Julian Date, and the Jews will ignore 23 Tevet, 5760 *(anno Mundi)*, and the Muslims will forget about 23 Ramadan, 1420 *(anno Hejirae)*. On second thought, probably not. They will all know that it is December 31, 1999 *(anno Domini)*, and there is something very special about the year 2000.

It's hard to say just why, but we humans love nice, round numbers with lots of zeros. How many of us remember being a child and going for a ride in a car that was about to top the 20,000-mile mark? Everybody in the car waits, silently, as 19,999.9 slowly creeps forward . . . and then, with a click, 20,000! All the children cheer.

December 31, 1999, is the evening when the great odometer in the sky clicks ahead.

The Zeroth Number

Waclaw Sierpinski, the great Polish mathematician . . . was worried that he'd lost one piece of his luggage. "No, dear!" said his wife. "All six pieces are here." "That can't be true," said Sierpinski, "I've counted them several times: zero, one, two, three, four, five."

— JOHN CONWAY AND RICHARD GUY,
THE BOOK OF NUMBERS

It may seem bizarre to suggest that Dionysius and Bede made a mistake when they forgot to include zero in their calendar.

After all, children count "one, two, three," not "zero, one, two." Except for the Mayans, nobody else had a year zero or started a month with day zero. It seems unnatural. On the other hand, when you count backward, it is second nature. Ten. Nine. Eight. Seven. Six. Five. Four. Three. Two. One. Liftoff.

The space shuttle always waits for zero before it blasts into the air. An important event happens at the *zero hour*, not the *one hour*. When you drive toward the site where a bomb went off, you're approaching *ground zero*.

If you look carefully enough, you will see that people usually *do* start counting with zero. A stopwatch starts ticking from 0:00.00 and only reaches 0:01.00 after a second has elapsed. A car's odometer comes from the factory set at 00000, though by the time the dealer's done tooling around town, it's probably got a few more miles on it. The military's day officially begins at 0000 hours. But count aloud and you always start with "one," unless you're a mathematician or a computer programmer.* It has to do with order.

When we are dealing with the *counting* numbers — 1, 2, 3, and so on — it is easy to rank them in order. One is the first counting number, two is the second counting number, and three is the third. We don't have to worry about mixing up the value of the number — its *cardinality* — with the order in which it arrives — its *ordinality* — since they are essentially the same thing. For years, this was the state of affairs, and everybody was happy. But as zero came into the fold, the neat relationship between a number's cardinality and its ordinality was ruined. The numbers went 0, 1, 2, 3: zero came first, one was second in line, and two was in third place. No longer

---

* When a computer programmer makes a program do something over and over, he'll more than likely make the computer count from, say, zero to nine to make the computer take ten steps. A forgetful programmer might make it count from one to nine, yielding only nine steps instead of ten. More than likely a bug like this was what ruined an Arizona lottery in 1998. In drawing after drawing, a nine never appeared. "They hadn't programmed it in," admitted a spokeswoman sheepishly.

were cardinality and ordinality interchangable. This is the root of the calendar problem.

The first hour of the day starts at zero seconds past midnight; the second hour starts at 1 AM, and the third hour starts at 2 AM. Though we count with the ordinals (first, second, third), we mark time with the cardinals (0, 1, 2). All of us have assimilated this way of thinking, whether we appreciate it or not. After a baby finishes his 12th month, we all say that the child is one year old; he has finished his first 12 months of life. If the baby turns one when she's already lived a year, isn't the only consistent choice to say that the baby is zero years old before that time? Of course, we say that the child is six weeks old or nine months old instead—a clever way of getting around the fact that the baby is zero.

Dionysius didn't have a zero, so he started the calendar with year 1, just as the ancients before him had started theirs. People of those times thought in terms of the old-style equivalence of cardinality and ordinality. That was just fine . . . for them. If zero never entered their minds, it could hardly be a problem.

The Gaping Void

> *It was not absolute nothingness. It was a kind of formlessness without any definition. . . . True reasoning convinced me that I should wholly subtract all remnants of every kind of form if I wished to conceive the absolutely formless. I could not achieve this.*
>
> —SAINT AUGUSTINE, *CONFESSIONS*

It's hard to blame the monks for their ignorance. The world of Dionysius Exiguus, Boethius, and Bede was dark indeed. Rome had collapsed, and Western civilization seemed but a shadow of Rome's past glory. The future seemed more horrid than the past. It is no wonder that in the search for wisdom medieval scholars didn't look to their peers for ideas. Instead, they turned to the ancients like Aristotle and the Neoplatonists. As these medieval thinkers imported the philosophy and science of the ancients, they inherited the ancient prejudices: a fear of the infinite and a horror of the void.

Medieval scholars branded void as evil—and evil as void. Satan was quite literally nothing. Boethius made the argument as follows: God is omnipotent. There is nothing God cannot do. But God, the ultimate goodness, cannot do evil. Therefore evil is nothing. It made perfect sense to the medieval mind.

Lurking underneath the veil of medieval philosophy, however, was a conflict. The Aristotelian system was Greek, but the Judeo-Christian story of creation was Semitic—and Semites didn't have such a fear of the void. The very act of creation was out of a chaotic void, and theologians like Saint Augustine, who lived in the fourth century, tried to explain it away by referring to the state before creation as "a nothing something" that is empty of form but yet "falls short of utter nothingness." The fear of the void was so great that Christian scholars tried to fix the Bible to match Aristotle rather than vice versa.

Luckily, not all civilizations were so afraid of zero.

Chapter 3
Nothing Ventured

[ZERO GOES EAST]

Where there is the Infinite there is joy. There is no joy in the finite.

— THE CHANDOGYA UPANISHAD

Though the West was afraid of the void, the East welcomed it. In Europe, zero was an outcast, but in India and later in the Arab lands, it flourished.

When we last saw zero, it was simply a placeholder. It was a blank spot in the Babylonian system of numeration. Zero was useful but was not truly a number on its own—it had no value. It only took its meaning from the digits to its left; the symbol for zero literally meant nothing if it was by itself. In India, all this changed.

In the fourth century BC, Alexander the Great marched with his Persian troops from Babylon to India. It was through this invasion that Indian mathematicians first learned about the Babylonian system of numbers—and about zero. When

Alexander died in 323 BC, his squabbling generals split his empire into pieces. Rome rose to power in the second century BC and swallowed up Greece, but Rome's power did not extend as far east as Alexander's had. As a result, remote India was insulated from the rise of Christianity and the fall of Rome in the fourth and fifth centuries AD.

India was also insulated from Aristotle's philosophy. Though Alexander had been tutored by Aristotle, and no doubt introduced India to Aristotelian ideas, the Greek philosophy never took hold. Unlike Greece, India never had a fear of the infinite or of the void. Indeed, it embraced them.

The void had an important place in the Hindu religion. Hinduism had started off as a polytheistic religion, a set of tales about warrior gods and battles similar in many ways to the Greek mythos. However, over centuries—centuries before Alexander arrived—the gods began to merge together. While Hinduism retained its popular rituals and devotion to its pantheon, at its core Hinduism became monotheistic and introspective. All the gods became aspects of an all-encompassing god, Brahma. At about the same time that the Greeks were rising in the Western world, Hinduism was becoming less like the Western myths; the individual gods became less distinct and the religion became more and more mystical. The mysticism was patently Eastern.

Like many Eastern religions, Hinduism was steeped in the symbolism of duality. (Of course, this idea occasionally came up in the Western world, where it was promptly branded as heretical. One example is the Manichaean heresy, which saw the world as being under the influence of equal and opposite sources of good and evil.) As with the yin and yang of the Far East and Zoroaster's dualism of good and evil in the Near East, creation and destruction were intermingled in Hinduism. The god Shiva was both creator and destroyer of the world and was depicted with the drum of creation in one hand and a flame of destruction in another (Figure 13). However, Shiva also represented nothingness. One aspect of

Figure 13: Shiva's dance

the deity, Nishkala Shiva, was literally the Shiva "without parts." He was the ultimate void, the supreme nothing—lifelessness incarnate. But out of the void, the universe was born, as was the infinite. Unlike the Western universe, the Hindu cosmos was infinite in extent; beyond our own universe were innumerable other universes.

At the same time, though, the cosmos never truly abandoned its original emptiness. Nothingness was what the world came from, and to achieve nothingness again became the ultimate goal of mankind. In one story, Death tells a disciple about the soul: "Concealed in the heart of all beings is the Atman, the Spirit, the Self," he says. "Smaller than the smallest atom, greater than the vast spaces." This Atman, which resides in every thing, is part of the essence of the universe, and is immortal. When a person dies, the Atman is released from the body and soon enters another being; the soul transmigrates and the person is reincarnated. The goal of the Hindu is to free the Atman entirely from the cycle of rebirth, to stop wandering from death to death. The way to achieve the ultimate liberation through lifelessness is to cease paying heed to the illusion of reality. "The body, the house of the spirit, is under the power of pleasure and pain," explains a god. "And if a man is ruled by his body then this man can never be free." But once you are able to separate yourself from the whims of the flesh and embrace the silence and nothingness of your soul, you will be liberated. Your Atman will fly from the web of human desire and join the collective consciousness—the infinite soul that suffuses the universe, at once everywhere and nowhere at the same time. It is infinity, and it is nothing.

So India, as a society that actively explored the void and the infinite, accepted zero.

Zero's Reincarnation

In the earliest age of the gods, existence was born from non-existence.

— THE RIG VEDA

Indian mathematicians did more than simply accept zero. They transformed it, changing its role from mere placeholder to number. This reincarnation was what gave zero its power.

The roots of Indian mathematics are hidden by time. An Indian text written the same year that Rome fell—476 AD— shows the influence of Greek, Egyptian, and Babylonian mathematics, brought by Alexander as he penetrated Indian lands. Like the Egyptians, the Indians had rope stretchers to

survey fields and lay out temples. They also had a sophisticated system of astronomy; like the Greeks, they tried to calculate the distance to the sun. That requires trigonometry; the
Indian version was probably derived from the system that the
Greeks had developed.

Sometime around the fifth century AD, Indian mathematicians changed their style of numbering; they moved from
a Greek-like system to a Babylonian-style one. An important
difference between the new Indian number system and the
Babylonian style was that Indian numbers were base-10 instead of base-60. Our numbers evolved from the symbols that
the Indians used; by rights they should be called Indian numerals rather than Arabic ones (Figure 14).

Nobody knows when the Indians made the switch to a
Babylonian-style place-value number system. The earliest
reference to the Hindu numerals comes from a Syrian bishop
who wrote, in 662, of how the Indians did calculations "by
means of nine signs." Nine—not ten. Zero was evidently not
among them. But it's hard to tell for sure. It is fairly clear that
the Hindu numerals were around before the bishop wrote
about them; there is evidence that zero appeared in some
variants of the Indian system by that time, though the bishop
hadn't heard about it. In any case, a symbol for zero—the
placeholder in the base-10 numbering system—was certainly
in use by the ninth century. By then Indian mathematicians
had already made a giant leap.

The Indians had borrowed little of Greek geometry. They
apparently didn't have a deep interest in the plane figures that
the Greeks loved so much. They never worried about
whether the diagonal of a square was rational or irrational,
nor did they investigate the conic sections as Archimedes
had. But they did learn how to play with numbers.

The Indian system of numbering allowed them to use
fancy tricks to add, subtract, multiply, and divide numbers
without using an abacus to help them. Thanks to their place-
number system, they could add and subtract large numbers in

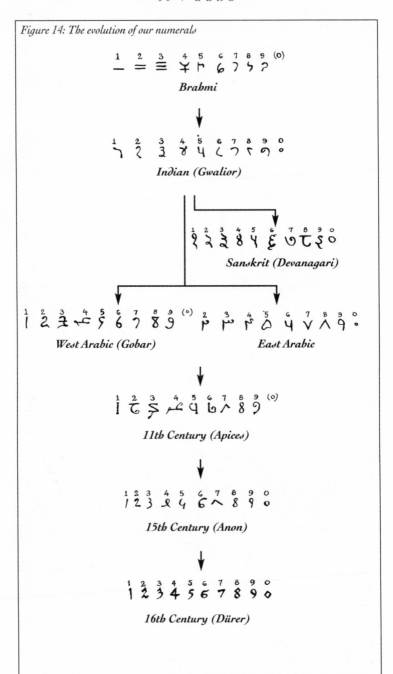

Figure 14: The evolution of our numerals

roughly the same way we do today. With training, a person could multiply with Indian numerals faster than an abacist could tally. Contests between the abacists and the so-called algorists who used Indian numerals were the medieval equivalents of the Kasparov versus Deep Blue chess match (Figure 15). Like Deep Blue, the algorists would win in the end.

Though the Indian number system was useful for everyday tasks like addition and multiplication, the true impact of Indian numbers was considerably deeper. Numbers had

Figure 15: An algorist versus an abacist

finally become distinct from geometry; numbers were used to do more than merely measure objects. Unlike the Greeks, the Indians did not see squares in square numbers or the areas of rectangles when they multiplied two different values. Instead, they saw the interplay of numerals—numbers stripped of their geometric significance. This was the birth of what we now know as *algebra*. Though this mind-set prevented the Indians from contributing much to geometry, it had another, unexpected effect. It freed the Indians from the shortcomings of the Greek system of thought—and their rejection of zero.

Once numbers shed their geometric significance, mathematicians no longer had to worry about mathematical operations making geometric sense. You can't remove a three-acre swath from a two-acre field, but nothing prevents you from subtracting three from two. Nowadays we recognize that $2 - 3 = -1$: negative one. However, this was not at all obvious to the ancients. Many times they solved equations only to get a negative result and concluded that their answer had no meaning. After all, if you are thinking in geometric terms, what is a negative area? It simply didn't make any sense to the Greeks.

To the Indians, negative numbers made perfect sense. Indeed, it was in India (and in China) that negative numbers first appeared. Brahmagupta, an Indian mathematician of the seventh century, gave rules for dividing numbers by each other, and he included the negatives. "Positive divided by positive, or negative by negative, is affirmative," he wrote. "Positive divided by negative is negative. Negative divided by affirmative is negative." These are the rules that we recognize today: divide two numbers and the answer is positive if the numbers' signs are the same.

Just as $2 - 3$ was now a number, so was $2 - 2$. It was zero. Not just a mere placeholder zero that represents an empty space on the abacus, but zero the number. It had a specific value, a fixed place on the number line. Since zero was equal to $2 - 2$, then it had to be placed between one $(2 - 1)$ and negative one $(2 - 3)$. Nothing else made sense. No longer could zero sit to the right of nine, just as it does on the top of the computer keyboard; zero had a position in the number line that was all its own. A number line without zero could no more exist than a number system without two. Zero had finally arrived.

However, even the Indians thought that zero was a pretty bizarre number, for all the usual reasons. After all, zero multiplied by anything is zero; it sucks everything into itself. And when you divide with it, all hell breaks loose. Brahmagupta tried to figure out what $0 \div 0$ and $1 \div 0$ were, and failed. "Cipher divided by cipher is naught," he wrote. "Positive or negative divided by cipher is a fraction with that for a denominator." In other words, he thought $0 \div 0$ was 0 (he was wrong, as we will see), and he thought that $1 \div 0$ was, well, we don't really know, because he doesn't make a whole lot of sense. Basically, he was waving his hands and hoping that the problem would go away.

Brahmagupta's mistake did not last for very long. In time the Indians realized that $1 \div 0$ was infinite. "This fraction of which the denominator is a cipher, is termed an infinite quantity," writes Bhaskara, a twelfth-century Indian mathematician, who tells of what happens when you add a number to $1 \div 0$. "There is no alteration, though many be inserted or extracted; as no change takes place in the infinite and immutable God."

God was found in infinity—and in zero.

The Arab Numeral

Does man forget that We created him out of the void?
— THE KORAN

By the seventh century, the West had withered with the fall of Rome, but the East was flourishing. India's growth was eclipsed by another Eastern civilization. As the star of the West sank below the horizon, another star was rising: Islam. Islam would take zero from India—and the West would eventually take it from Islam. Zero's rise to preeminence had to begin in the East.

One evening in 610 AD, Mohammed, a thirty-year-old native of Mecca, fell into a trance on Mount Hira. According to legend, the angel Gabriel told him, "Recite!" Mohammed did, and his divine revelations started a wildfire. A decade after Mohammed's death in 632, his followers had captured Egypt, Syria, Mesopotamia, and Persia. Jerusalem, the holy city of the Jews and the Christians, had fallen. By 700, Islam stretched as far as the Indus River in the East and Algiers in the West. In 711 the Muslims captured Spain, and they advanced as far as France. In the East they defeated the Chinese in 751. Their empire stretched farther than even Alexander could have imagined. Along the way to China, the Muslims conquered India. And there the Arabs learned about Indian numerals.

The Muslims were quick to absorb the wisdom of the peoples that they conquered. Scholars started translating texts into Arabic, and in the ninth century Caliph al-Mamun founded a great library: the House of Wisdom at Baghdad. It was to become the center of learning in the Eastern world—and one of its first scholars was the mathematician Mohammed ibn-Musa al-Khowarizmi.

Al-Khowarizmi wrote several important books, like *Al-jabr wa'l muqabala*, a treatise on how to solve elementary equations; the *Al-jabr* in the title (which means something like "completion") gave us the term algebra. He also wrote a book about the Hindu numeral system, which allowed the new style of numbers to spread quickly through the Arab world— along with *algorithms*, the tricks for multiplying and dividing Hindu numerals quickly. In fact, the word *algorithm* was a corruption of al-Khowarizmi's name. Though the Arabs took the notation from India, the rest of the world would dub the new system Arabic numerals.

The very word *zero* smacks of its Hindu and Arabic roots. When the Arabs adopted Hindu-Arabic numerals, they also adopted zero. The Indian name for zero was *ʃunya,* meaning "empty," which the Arabs turned into *ʃifr.* When some Western scholars described the new number to their colleagues, they turned *ʃifr* into a Latin-sounding word, yielding *zephirus,* which is the root of our word zero. Other Western mathematicians didn't change the word so heavily and called zero *cifra,* which became *cipher.* Zero was so important to the new set of numbers that people started calling all numbers ciphers, which gave the French their term *chiffre,* digit.

However, when al-Khowarizmi was writing about the Hindu system of numbers, the West was still far from adopting zero. Even the Muslim world, with its Eastern traditions, was heavily contaminated by the teachings of Aristotle, thanks to the conquests of Alexander the Great. However, as Indian mathematicians had made quite clear, zero was the embodiment of the void. Thus, if the Muslims were to accept zero, they had to reject Aristotle. That was precisely what they did.

A twelfth-century Jewish scholar, Moses Maimonides, described the Kalam—the beliefs of Islamic theologians—with horror. He noted that instead of accepting Aristotle's proof of God, the Muslim scholars turned to the atomists, Aristotle's old rivals, whose doctrine, though out of favor, managed to survive the ravages of time. The atomists, remember, held that matter was composed of individual particles called atoms, and if these particles were able to move about, there had to be a vacuum between them, otherwise the atoms would be bumping into one another, unable to get out of one another's way.

The Muslims seized upon the atomists' ideas; after all, now that zero was around, the void was again a respectable idea. Aristotle hated the void; the atomists required it. The Bible told of the creation from the void, while the Greek doctrine rejected the possibility. The Christians, cowed by the power of Greek philosophy, chose Aristotle over their Bible. The Muslims, on the other hand, made the opposite choice.

I Am That I Am: Nothing

Nothingness is being and being nothingness.... Our limited mind can not grasp or fathom this, for it joins infinity.

— AZRAEL OF GERONA

Zero was an emblem of the new teachings, of the rejection of Aristotle and the acceptance of the void and the infinite. As Islam spread, zero diffused throughout the Muslim-controlled world, everywhere conflicting with Aristotle's doctrine. Islamic scholars battled back and forth, and in the eleventh century a Muslim philosopher, Abu Hamid al-Ghazali, declared that clinging to Aristotelian doctrine should be punishable by death. The debate ended shortly thereafter.

It's no surprise that zero caused such discord. The Muslims, with their Semitic, Eastern background, believed that God created the universe out of the void — a doctrine that could never be accepted where people shared Aristotle's hatred of the void and of the infinite. As zero spread through the Arab lands, the Muslims embraced it and rejected Aristotle. The Jews were the next in line.

For millennia the center of Jewish life had been planted firmly in the Middle East, but in the tenth century an opportunity for Jews arose in Spain. Caliph Abd al-Rahman III had a Jewish minister who imported a number of intellectuals from Babylonia. Soon a large Jewish community flourished in Islamic Spain.

Early medieval Jews, both in Spain and in Babylon, were wed firmly to Aristotle's doctrines. Like their Christian counterparts, they refused to believe in the infinite or the void. However, just as Aristotelian philosophy conflicted with Islamic teachings, it conflicted with Jewish theology. This is what drove Maimonides, the twelfth-century rabbi, to write a tome to reconcile the Semitic, Eastern Bible with the Greek, Western philosophy that permeated Europe.

From Aristotle, Maimonides had learned to prove God's existence by denying the infinite. Reproducing the Greek arguments faithfully, Maimonides contended that the hollow spheres that twirled about the earth had to be moved by something, say, the next sphere out. But the next sphere out had to be moved by something—the next sphere in line. However, since there cannot be an infinite number of spheres (because infinity was impossible), something had to be moving the outermost sphere. That was the prime mover: God.

Maimonides' argument was, indeed, a "proof" of God's existence—something incredibly valuable in any theology. Yet at the same time, the Bible and other Semitic traditions were full of the ideas of the infinite and the void, ideas that the Muslims already embraced. Just like Saint Augustine 800 years earlier, Maimonides tried to reshape the Semitic Bible to fit into Greek doctrine: doctrine that had an unreasonable fear of the void. But unlike the early Christians, who had freed themselves to interpret parts of the Old Testament as metaphor, Maimonides was unwilling to Hellenize his religion completely. Rabbinic tradition compelled him to accept the biblical account of the universe's creation from the void. This, in turn, meant contradicting Aristotle.

Maimonides argued that there were flaws in Aristotle's proof that the universe had always existed. After all, it conflicted with the Scriptures. This, of course, meant that Aristotle had to go. Maimonides stated that the act of creation came from nothing. It was *creatio ex nihilo,* despite the Aristotelian ban on the vacuum. With that stroke the void moved from sacrilege to holiness.

For the Jews, the years after Maimonides' death became the era of nothing. In the thirteenth century a new doctrine spread: kabbalism, or Jewish mysticism. One centerpiece of kabbalistic thought is gematria—the search for coded messages within the text of the Bible. Like the Greeks, the

Hebrews used letters from their alphabet to represent numbers, so every word had a numerical value. This could be used to interpret the hidden meaning of words. For instance, Gulf War participants might have noticed that Saddam has the following value: samech (60) + aleph (1) + daled (4) + aleph (1) + mem (600) = 666—a number that Christians associate with the evil Beast that appears during the Apocalypse. (Whether "Saddam" has two daleds or one would make no difference to the kabbalists, who often used alternate spellings of words to make sums come out right.) Kabbalists thought that words and phrases with the same numerical value were mystically linked. For instance, Genesis 49:10 states, "The scepter shall not depart from Judah . . . until Shiloh come." The Hebrew phrase for "until Shiloh come" has a value of 358, exactly the same for the Hebrew word *meshiach*, messiah. Hence, the passage presages the coming of the Messiah. Certain numbers were holy or evil, according to the kabbalists—and they looked through the Bible for these numbers and for hidden messages found by scanning through it in various ways. A recent bestseller, *The Bible Code*, purported to find prophecies by this method.

The kabbalah was much more than number crunching; it was a tradition so mystical that some scholars say that it bears a striking resemblance to Hinduism. For instance, the kabbalah seized upon the idea of the dual nature of God. The Hebrew term *ein sof*, which meant "infinite," represented the creator aspect of God, the part of the deity that made the universe and that permeates every corner of the cosmos. But at the same time it had a different name: *ayin*, or "nothing." The infinite and the void go hand in hand, and are both part of the divine creator. Better yet, the term *ayin* is an anagram of (and has the same numerical value as) the word *aniy*, the Hebrew "I." It could scarcely be clearer: God was saying, in code, "I am nothing." And at the same time, infinity.

As the Jews pitted their Western sensibilities against their Eastern Bible, the same battle was under way in the Christian world. Even as the Christians battled the Muslims—during

Charlemagne's reign in the ninth century and during the Crusades in the eleventh, twelfth, and thirteenth centuries — warrior-monks, scholars, and traders began to bring Islamic ideas back to the West. Monks discovered that the astrolabe, an Arabic invention, was a handy tool for keeping track of time in the evening, helping them keep their prayers on schedule. The astrolabes were often inscribed with Arabic numerals.

The new numbers didn't catch on, even though a tenth-century pope, Sylvester II, was an admirer of them. He probably learned about the numerals during a visit to Spain and brought them back with him when he returned to Italy. But the version he learned did not have a zero — and the system would have been even less popular if it had. Aristotle still had a firm grip on the church, and its finest thinkers still rejected the infinitely large, the infinitely small, and the void. Even as the Crusades drew to a close in the thirteenth century, Saint Thomas Aquinas declared that God could not make something that was infinite any more than he could make a scholarly horse. But that implied that God was not omnipotent — a forbidden thought in Christian theology.

In 1277 the bishop of Paris, Étienne Tempier, called an assembly of scholars to discuss Aristotelianism, or rather, to attack it. Tempier abolished many Aristotelian doctrines that contradicted God's omnipotence, such as, "God can not move the heavens in a straight line, because that would leave behind a vacuum." (The rotating spheres caused no problem, because they still occupied the same space. It is only when you move the spheres in a line that you are forced to have a space to move the heavens into, and you are forced to have a space behind them after they move.) God could make a vacuum if he wanted. All of a sudden the void was allowed, because an omnipotent deity doesn't need to follow Aristotle's rules if he doesn't want to.

Tempier's pronouncements were not the final blow to Aristotelian philosophy, but they certainly signaled that the foundations were crumbling. The church would cling to Aristotle for a few more centuries, but the fall of Aristotle and the rise of the void and the infinite were clearly beginning. It was a propitious time for zero to arrive in the West. In the mid-twelfth century the first adaptations of al-Khowarizmi's *Al-jabr* were working their way through Spain, England, and the rest of Europe. Zero was on the way, and just as the church was breaking the shackles of Aristotelianism, it arrived.

Zero's Triumph

. . . a profound and important idea which appears so simple to us now that we ignore its true merit. But its very simplicity and the great ease which it lent to all computations put our arithmetic in the first rank of useful inventions.

— PIERRE-SIMON LAPLACE

Christianity initially rejected zero, but trade would soon demand it. The man who reintroduced zero to the West was Leonardo of Pisa. The son of an Italian trader, he traveled to northern Africa. There the young man—better known as Fibonacci—learned mathematics from the Muslims and soon became a good mathematician in his own right.

Fibonacci is best remembered for a silly little problem he posed in his book, *Liber Abaci,* which was published in 1202. Imagine that a farmer has a pair of baby rabbits. Babies take two months to reach maturity, and from then on they produce another pair of rabbits at the beginning of every month. As these rabbits mature and reproduce, and those rabbits mature and reproduce, and so on, how many pairs of rabbits do you have during any given month?

Well, during the first month, you have one pair of rabbits, and since they haven't matured, they can't reproduce.

During the second month you still have only one pair.

But at the beginning of the third month, the first pair reproduces: you've got two pairs.

At the beginning of the fourth month, the first pair reproduces again, but the second pair is not mature enough: three pairs.

The next month the first pair reproduces, the second pair reproduces, since it has reached maturity, but the third pair is too young. That is two additional pairs of rabbits: five in all.

The number of rabbits goes as follows: 1, 1, 2, 3, 5, 8, 13, 21, 34, 55, . . . ; the number of rabbits you have in any given month is the sum of the rabbits that you had in each of the two previous months. Mathematicians instantly realized the importance of this series. Take any term and divide it by its previous term. For instance, $8/5 = 1.6$; $13/8 = 1.625$; $21/13 = 1.61538$ . . . . These ratios approach a particularly interesting number: the golden ratio, which is 1.61803 . . . .

Pythagoras had noticed that nature seemed to be governed by the golden ratio. Fibonacci discovered the sequence that is responsible. The size of the chambers of the nautilus and the number of clockwise grooves to counterclockwise grooves in the pineapple are governed by this sequence. This is why their ratios approach the golden ratio.

Though this sequence is the source of Fibonacci's fame, Fibonacci's *Liber Abaci* had a much more important purpose than animal husbandry. Fibonacci had learned his mathematics from the Muslims, so he knew about Arabic numerals, including zero. He included the new system in *Liber Abaci*, finally introducing Europe to zero. The book showed how useful Arabic numerals were for doing complex calculations, and the Italian merchants and bankers quickly seized upon the new system, zero included.

Before Arabic numerals came around, money counters had to make do with an abacus or a counting board. The Germans called the counting board a *Rechenbank*, which is why we call moneylenders *banks*. At that time, banking methods were primitive. Not only did they use counting

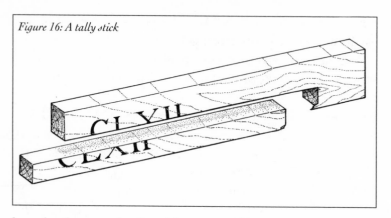

Figure 16: A tally stick

boards, they used *tally sticks* to record loans: a money value was written along the stick's side, and it was split in two (Figure 16). The lender kept the biggest piece, the *stock*. After all, he was the stockholder.[*]

Italian merchants loved the Arabic numbers. They allowed the bankers to get rid of their counting boards. However, while businessmen saw their usefulness, the local governments hated them. In 1299, Florence banned Arabic numerals. The ostensible reason was that the numbers were easily changed and falsified. (A 0 could be turned into a 6 with a simple flourish of a pen, for instance.) But the advan-

---

[*] Tally sticks caused no end of trouble. The English Exchequer used to keep accounts on a variant of the tally stick until 1826. Charles Dickens told of the outcome of that long-outdated practice: "In 1834, it was found that there was a considerable accumulation of them; and the question then arose, what was to be done with such worn-out, worm-eaten, rotten old bits of wood? The sticks were housed in Westminster, and it would naturally occur to any intelligent person that nothing could be easier than to allow them to be carried away for firewood by the miserable people who lived in that neighborhood. However, they never had been useful, and official routine required that they should never be, and so the order went out that they were to be privately and confidentially burned. It came to pass that they were burned in a stove in the House of Lords. The stove, over-gorged with these preposterous sticks, set fire to the panelling; the panelling set fire to the House of Commons; the two houses were reduced to ashes; architects were called in to build others; and we are now in the second million of the cost thereof."

tages of zero and the other Arabic numerals were not so easily dispensed with; Italian merchants continued to use them, and even used them to send encrypted messages—which is how the word *cipher* came to mean "secret code."

In the end the governments had to relent in the face of commercial pressure. The Arabic notation was allowed into Italy and soon spread throughout Europe. Zero had arrived—as had the void. The Aristotelian wall was crumbling, thanks to the influence of the Muslims and the Hindus, and by the 1400s even the staunchest European supporters of Aristotelianism had their doubts. Thomas Bradwardine, who was to become archbishop of Canterbury, tried to disprove atomism, Aristotle's old nemesis. At the same time, he wondered whether his own logic was faulty, since he based his arguments on geometry, whose infinitely divisible lines automatically reject atomism. However, the battle against Aristotle was far from over. If Aristotle were to fall, the proof of God—a bulwark of the church—was no longer valid. A new proof was needed.

Worse yet, if the universe were infinite, then there could be no center. How could Earth, then, be the center of the universe? The answer was found in zero.

Chapter **4**
The Infinite God of Nothing

[THE THEOLOGY OF ZERO]

And new philosophy calls all in doubt,

The element of fire is quite put out;

The sun is lost, and th' earth, and no man's wit

Can well direct him where to look for it. . . .

'Tis all in pieces, all coherence gone;

All just supply, and all relation:

Prince, subject, Father, Son, are things forgot.

— JOHN DONNE,
"AN ANATOMY OF THE WORLD"

Zero and infinity were at the very center of the Renaissance. As Europe slowly awakened from the Dark Ages, the void and the infinite—nothing and everything—would destroy the Aristotelian foundation of the church and open the way to the scientific revolution.

At first the papacy was blind to the danger. High-ranking clergymen experimented with the dangerous ideas of the void and the infinite, even though the ideas struck at the core of the ancient Greek philosophy that the church cherished so much. Zero appeared in the middle of every Renaissance painting, and a cardinal declared that the universe was infinite—boundless. However, the brief love affair with zero and infinity was not to last.

When the church was threatened, it retreated into its old philosophy, turning back toward the Aristotelian doctrine that had supported it for so many years. It was too late. Zero had taken hold in the West, and despite the papacy's objections, it was too strong to be exiled once more. Aristotle fell to the infinite and to the void, and so did the proof of God's existence.

Only one option remained for the church: accept zero and infinity. Indeed, to the devout, God could be found, hidden within the void and the infinite.

The Nutshell Cracked

O God, I could be bounded in a nutshell and count myself a king of infinite space, were it not that I have bad dreams.

—WILLIAM SHAKESPEARE, *HAMLET*

At the beginning of the Renaissance, it was not obvious that zero would pose a threat to the church. It was an artistic tool, an infinite nothing that ushered in the great Renaissance in the visual arts.

Before the fifteenth century, paintings and drawings were largely flat and lifeless. The images in them were distorted and two-dimensional; gigantic, flat knights peered out of tiny, misshapen castles (Figure 17). Even the best artists could not draw a realistic scene. They did not know how to use the power of zero.

Figure 17: Flat knights and misshapen castles

It was an Italian architect, Filippo Brunelleschi, who first demonstrated the power of an infinite zero: he created a realistic painting by using a vanishing point.

By definition, a point is a zero—thanks to the concept of dimension. In everyday life you deal with three-dimensional objects. (Actually, Einstein revealed that our world is four-dimensional, as we shall see in a later chapter.) The clock on your dresser, the cup of coffee you drink in the morning, the book you're reading right now—all these are three-dimensional objects. Now imagine that a giant hand comes down and squashes the book flat. Instead of being a three-dimensional object, the book is now a flat, floppy rectangle. It has lost a dimension; it has length and width, but no height. It is now two-dimensional. Now imagine that the book, turned sideways, is crushed once again by the giant hand. The book is no longer a rectangle. It is a line. Again, it has lost a dimension; it has neither height nor width, but it has length. It is a one-dimensional object. You can take away even this single dimension. Squashed along its length, the line becomes a point, an infinitesimal nothing with no length, no width, and no height. A point is a zero-dimensional object.

In 1425, Brunelleschi placed just such a point in the center of a drawing of a famous Florentine building, the Baptistery. This zero-dimensional object, the vanishing point, is an infinitesimal dot on the canvas that represents a spot infinitely far away from the viewer (Figure 18). As objects recede into the distance in the painting, they get closer and closer to the vanishing point, getting more compressed as they get farther away from the viewer. Everything sufficiently distant—people, trees, buildings—is squashed into a zero-dimensional point and disappears. The zero in the center of the painting contains an infinity of space.

This apparently contradictory object turned Brunelleschi's drawing, almost magically, into such a good likeness of the three-dimensional Baptistery building that it was indistin-

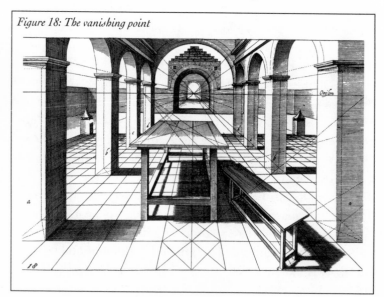

Figure 18: The vanishing point

guishable from the real thing. Indeed, when Brunelleschi used a mirror to compare the painting and the building, the reflected image matched the building's geometry exactly. The vanishing point turned a two-dimensional drawing into a perfect simulation of a three-dimensional building.

It is no coincidence that zero and infinity are linked in the vanishing point. Just as multiplying by zero causes the number line to collapse into a point, the vanishing point has caused most of the universe to sit in a tiny dot. This is a *singularity*, a concept that became very important later in the history of science—but at this early stage, mathematicians knew little more than the artists about the properties of zero. In fact, in the fifteenth century, artists were amateur mathematicians. Leonardo da Vinci wrote a guide to drawing in perspective. Another of his books, about painting, warns, "Let no one who is not a mathematician read my works." These mathematician-artists perfected the technique of perspective and could soon depict arbitrary objects in three dimensions. No longer would artists be restricted to flat likenesses. Zero had transformed the art world.

Zero was, quite literally, at the center of Brunelleschi's painting. The church, too, dabbled with zero and the infinite, though church doctrine was still dependent on Aristotelian ideas. A contemporary of Brunelleschi, a German cardinal named Nicholas of Cusa, looked at infinity and promptly declared, "Terra non est centra mundi": the earth is not the center of the universe. The church didn't yet realize how dangerous, how revolutionary, that idea was.

One of the old declarations of the medieval Aristotelian doctrine—as strong as the ban on the vacuum—was the statement that Earth was unique. It was at the universe's very center. Earth's special position at the center of the universe made it the only world capable of containing life, as Aristotle held that all objects sought out their proper place. Heavy objects, like rocks or people, belonged on the ground; light objects, like air, belonged in the heavens. Not only did this imply that the planets—in the heavens—were made of light, airy stuff, but it also meant that any people in the heavens would naturally fall to Earth. Thus creatures could only inhabit the nutmeat in the center of the nutshell cosmos. Having other planets with life on them was as silly as having a sphere with two centers.

When Tempier declared that the omnipotent God could create a vacuum if he so desired, Tempier insisted that God could break any Aristotelian law. God could create life on other worlds if he wished. There could be thousands of other Earths, each teeming with creatures; it was certainly within God's power, whether Aristotle agrees or not.

Nicholas of Cusa was bold enough to say that God *must* have done so. "The regions of the other stars are similar to this," he said, "for we believe that none of them is deprived of inhabitants." The sky was littered with an infinite number of stars. The planets glowed in the heavens; the moon and the sun each glowed with light. Why couldn't the stars in the sky be planets or moons or suns on their own? Maybe Earth glows brightly in their heavens, just as they glow in ours. Nicholas was sure that God had, indeed, created an infinite number of other worlds. Earth was no longer at the center of the universe. Yet Nicholas was not declared a heretic, and the church didn't react to the new idea.

In the meantime another Nicholas turned Cusa's philosophy into a scientific theory. Nicolaus Copernicus showed that Earth is not the center of the universe. It revolves around the sun.

A Polish monk and a physician, Copernicus learned mathematics so he could cast astrological tables, the better to cure his patients with. Along the way, Copernicus's dabblings with the planets and stars showed him how complicated the old Greek system of tracking the planets was. Ptolemy's clockwork heavens—with Earth at the center—were extremely accurate. However, they were terribly complex. Planets course around the sky throughout the year, but every so often they stop, move backward, and then shoot ahead once more. To account for the planets' bizarre behavior,

Ptolemy added *epicycles* to his planetary clockwork: little circles within circles could explain the backward, or *retrograde*, motion of the planets (Figure 19).

The power of Copernicus's idea was in its simplicity. Instead of placing Earth at the center of the universe filled with epicycle-filled clockworks, Copernicus imagined that the sun was at the center instead, and the planets moved in simple circles. Planets would seem to zoom backward as Earth overtook them; no epicycles were needed. Though Copernicus's system didn't agree with the data completely—the circular orbits were wrong, though the heliocentric idea was correct—it was much simpler than the Ptolemaic system. The earth revolved around the sun. *Terra non est centra mundi.*

Nicholas of Cusa and Nicolaus Copernicus cracked open the nutshell universe of Aristotle and Ptolemy. No longer was the earth comfortably ensconced in the center of the universe; there was no shell containing the cosmos. The universe went on into infinity, dotted with innumerable worlds, each inhabited by mysterious creatures. But how could Rome claim to be the seat of the one true Church if its authority could not extend to other solar systems? Were there other popes on other planets? It was a grim prospect for the Catholic Church, especially since it was beginning to have trouble with its subjects on even its own world.

Copernicus published his magnum opus on his deathbed—in 1543, just before the church started clamping down on new ideas. Copernicus's book, *De Revolutionibus*, was even dedicated to Pope Paul III. However, the church was under attack. As a result, the new ideas—the questioning of Aristotle—could no longer be tolerated.

The attack on the church began in earnest in 1517, when a constipated German monk nailed a list of complaints to the door of the church in Wittenberg. (Luther's constipation was legendary. Some scholars believe that his great revelation about faith came to him when he was sitting on the privy. "Luther's release from the constricting bondage of fear corresponded to the release of his bowels," notes one text, com-

Figure 19: Epicycles, retrograde motion, and heliocentrism

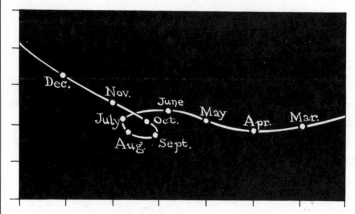

Retrograde Motion of Mars (An actual track)

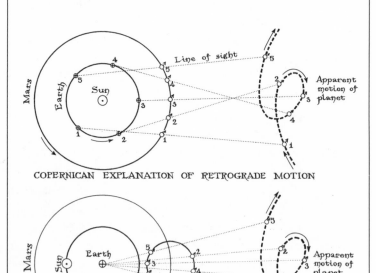

COPERNICAN EXPLANATION OF RETROGRADE MOTION

PTOLEMAIC EXPLANATION OF RETROGRADE MOTION

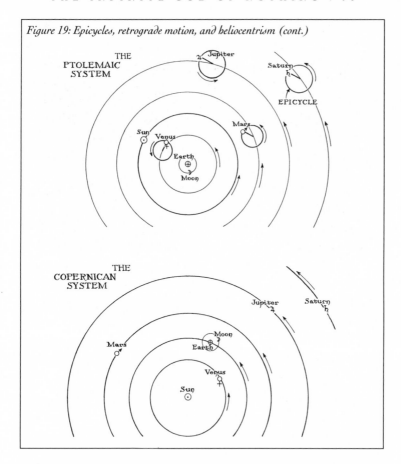

Figure 19: Epicycles, retrograde motion, and heliocentrism (cont.)

menting on this theory.) This was the beginning of the Refor-
mation; intellectuals everywhere began to reject the authority
of the pope. By the 1530s, in a quest to ensure an orderly suc-
cession to the throne, Henry VIII had spurned the authority
of the pope, declaring himself the head of the English clergy.

The Catholic Church had to strike back. Though it
had been experimenting with other philosophies for several
centuries, when threatened with schism it turned orthodox
once again. It fell back upon its orthodox teachings—the
Aristotelian-based philosophies of scholars like Saint Augustine and Boethius, as well as Aristotle's proof of God. No
longer could cardinals and clerics question the ancient doctrines. Zero was a heretic. The nutshell universe had to be ac-

cepted; the void and the infinite must be rejected. One of the key groups that spread these teachings was founded in the 1530s: the Jesuit order, a collection of highly trained intellectuals well suited to attack Protestantism. The church had other tools to fight heresy as well; the Spanish Inquisition started burning Protestants in 1543, the same year Copernicus died and the same year that Pope Paul III issued the Index of forbidden books. The Counter-Reformation was the church's attempt to rebuild the old order by crushing the new ideas. An idea embraced by Bishop Étienne Tempier in the thirteenth century and Cardinal Nicholas of Cusa in the fifteenth century could mean a death sentence in the sixteenth century.

This is what happened to the unfortunate Giordano Bruno. In the 1580s, Bruno, a former Dominican cleric, published *On the Infinite Universe and Worlds*, where he suggested, like Nicholas of Cusa, that the earth was not the center of the universe and that there were infinite worlds like our own. In 1600 he was burned at the stake. In 1616 the famous Galileo Galilei, another Copernican, was ordered by the church to cease his scientific investigations. The same year, Copernicus's *De Revolutionibus* was placed on the Index of forbidden books. An attack on Aristotle was considered an attack upon the church.

Despite the church's Counter-Reformation, the new philosophy wasn't easily destroyed. It got stronger as time went on, thanks to the investigations of Copernicus's successors. In the beginning of the seventeenth century, another astrologer-monk, Johannes Kepler, refined Copernicus's theory, making it even more accurate than the Ptolemaic system. Instead of moving in circles, the planets, including Earth, moved in ellipses around the sun. This explained the motion of the planets in the heavens with incredible accuracy; no longer could astronomers object that the heliocentric system was inferior to the geocentric one. Kepler's model was simpler than Ptolemy's, and it was more accurate. Despite the church's objections, Kepler's heliocentric system would prevail eventually, because Kepler was right and Aristotle was wrong.

The church attempted to patch the holes in the old way of thought, but Aristotle, the geocentric world, and the feudal way of life were all mortally wounded. Everything that philosophers had taken for granted for millennia was called into doubt. The Aristotelian system could not be trusted, and at the same time it could not be rejected. What, then, could be taken for granted? Literally nothing.

Zero and the Void

I am in a sense something intermediate between God and

nought.

— RENÉ DESCARTES, *DISCOURSE ON METHOD*

Zero and the infinite were at the very center of the philosophical war taking place during the sixteenth and seventeenth centuries. The void had weakened Aristotle's philosophy, and the idea of an infinitely large cosmos helped shatter the nutshell universe. The earth could not be at the center of God's creation. As the papacy lost its hold on its flock, the Catholic Church tried to reject zero and the void more strongly than ever, yet zero had already taken root. Even the most devout intellectuals—the Jesuits—were torn between the old, Aristotelian ways and the new philosophies that included zero and the void, infinity and the infinite.

René Descartes was trained as a Jesuit, and he, too, was torn between the old and the new. He rejected the void but put it at the center of his world. Born in 1596 in the center of France, Descartes would bring zero to the center of the number line, and he would seek a proof of God in the void and the infinite. Yet Descartes could not reject Aristotle entirely; he was so afraid of the void that he denied its existence.

Like Pythagoras, Descartes was a mathematician-philosopher; perhaps his most lasting legacy was a mathematical invention—what we now call Cartesian coordinates. Anyone who has taken geometry in high school has seen them: they are the sets of numbers in parentheses that represent a point in space. For instance, the symbol (4, 2) represents a point four units to the right and two units upward. But to the right and upward of what? The Origin. Zero (Figure 20).

Descartes realized that he could not start his two reference lines, or axes, with the number 1. That would lead to an error like the one Bede encountered when revamping the calendar. However, unlike Bede, he lived in a Europe where Arabic numerals were common, so he started counting with zero. At the very center of the coordinate system—where the two axes cross—sits a zero. The origin, the point (0, 0), is the foundation of the Cartesian system of coordinates. (Descartes's

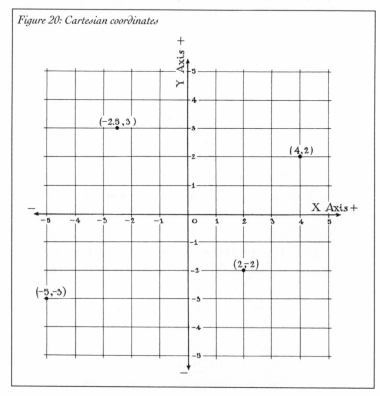

Figure 20: Cartesian coordinates

notation was slightly different from what we use today. For one thing, he didn't extend his coordinate system to the negative numbers, though his colleagues would soon do that for him.)

Descartes quickly realized how powerful his coordinate system was. He used it to turn figures and shapes into equations and numbers; with Cartesian coordinates every geometric object—squares, triangles, wavy lines—could be represented by an equation, a mathematical relationship. For example, a circle at the origin can be represented by the set of all points where $x^2 + y^2 - 1 = 0$. A parabola might be $y - x^2 = 0$. Descartes unified numbers and shapes. No longer were the Western art of geometry and the Eastern art of algebra separate domains. They were the same thing, as every shape could simply be expressed as an equation of the form $f(x,y) = 0$ (Figure 21). Zero was at the center of the coordinate system, and zero was implicit in each geometric shape.

To Descartes, zero was also implicit in God's domain, as was the infinite. Since the old Aristotelian doctrine was crumbling, Descartes, true to his Jesuit training, tried to use nought and infinity to replace the old proof of God's existence.

Like the ancients, Descartes assumed that nothing, not even knowledge, can be created out of nothing, which means that all ideas—all philosophies, all notions, all future discoveries—already exist in people's brains when they are born. Learning is just the process of uncovering that previously imprinted code of laws about the workings of the universe. Since we have a concept of an infinite perfect being in our minds, Descartes then argued that this infinite and perfect being—God—must exist. All other beings are less than divine; they are finite. They all lie somewhere between God and nought. They are a combination of infinity and zero.

However, though zero appeared and reappeared throughout Descartes's philosophy, Descartes insisted unto his death that the void—the ultimate zero—does not exist. A child of

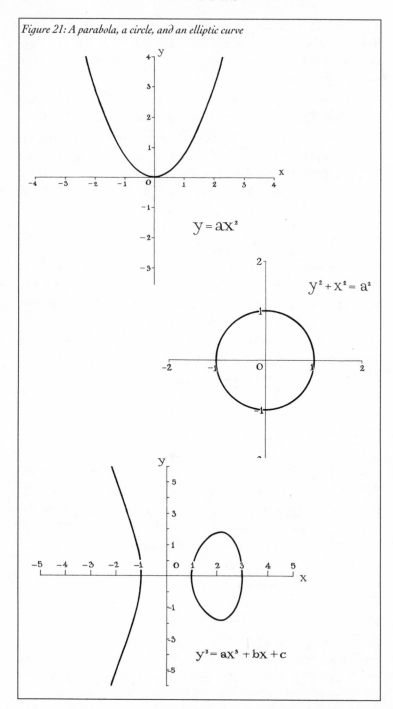

Figure 21: A parabola, a circle, and an elliptic curve

$y = ax^2$

$y^2 + x^2 = a^2$

$y^2 = ax^3 + bx + c$

the Counter-Reformation, Descartes learned about Aristotle at the very moment when the church was relying upon his principles the most. As a result, Descartes, indoctrinated with the Aristotelian philosophy, denied the existence of the vacuum.

It was a difficult position to take; Descartes was certainly mindful of the metaphysical problems of rejecting the vacuum entirely. Later in his life he wrote about atoms and the vacuum: "About these things that involve a contradiction, it can absolutely be said that they cannot happen. However, one shouldn't deny that they can be done by God, namely, if he were to change the laws of nature." Yet, like the medieval scholars before him, Descartes believed that nothing truly moved in a straight line, for that would leave a vacuum behind it. Instead, everything in the universe moved in a circular path. It was a truly Aristotelian way of thinking—yet the void would soon unseat Aristotle once and for all.

Even today, children are taught "Nature abhors a vacuum," while the teachers don't really understand where that phrase came from. It was an extension of the Aristotelian philosophy: vacuums don't exist. If someone would attempt to create a vacuum, nature would do anything in its power to prevent it from happening. It was Galileo's secretary, Evangelista Torricelli, who proved that this wasn't true—by creating the first vacuum.

In Italy workmen used a kind of pump, which worked more or less like a giant syringe, to raise water out of wells and canals. This pump had a piston that fitted snugly in a tube. The bottom of the tube was placed in the water, so when the piston was raised, the water level followed the plunger upward.

Galileo heard from a worker that these pumps had a problem: they could only lift water about 33 feet. After that, the plunger kept on going upward, yet the water level stayed the same. It was a curious phenomenon, and Galileo passed the problem on to his assistant, Torricelli, who started doing experiments, trying to figure out the reason for the pumps' curious limitation. In 1643, Torricelli took a long tube that was closed at one end and filled it with mercury. He upended it, placing the open end in a dish also filled with mercury. Now, if Torricelli had upended the tube in air, everyone would expect the mercury to run out, because it would quickly be replaced by air; no vacuum would be created. But when it was upended in a dish of mercury, there was no air to replace the mercury in the tube. If nature truly abhorred a vacuum so much, the mercury in the tube would have to stay put so as not to create a void. The mercury didn't stay put. It sank downward a bit, leaving a space at the top. What was in that space? Nothing. It was the first time in history anyone had created a sustained vacuum.

No matter the dimensions of the tube that Torricelli used, the mercury would sink down until its highest point was about 30 inches above the dish; or, looking at it another way, mercury could only rise 30 inches to combat the vacuum above it. Nature only abhorred a vacuum as far as 30 inches. It would take an anti-Descartes to explain why.

In 1623, Descartes was twenty-seven, and Blaise Pascal, who would become Descartes's opponent, was zero years old. Pascal's father, Étienne, was an accomplished scientist and mathematician; the young Blaise was a genius equal to his father. As a young man, Blaise invented a mechanical calculating machine, named the Pascaline, which is similar to some of the mechanical calculators that engineers used before the invention of the electronic calculator.

When Blaise was twenty-three, his father slipped on a patch of ice and broke his thigh. He was cared for by a group of Jansenists, Catholics who belonged to a sect based largely on a hatred of the Jesuit order. Soon the entire Pascal family was won over, and Blaise became an anti-Jesuit, a counter-counter-reformationist. Pascal's newfound religion was not a comfortable fit for the young scientist. Bishop Jansen, the founder of the sect, had declared that science was sinful; curiosity about the natural world was akin to lust. Luckily, Pascal's lust was greater than his religious fervor for a time, because he would use science to unravel the secret of the vacuum.

About the time of the Pascals' conversion, a friend of Étienne's—a military engineer—came to visit and repeated Torricelli's experiment for the Pascals. Blaise Pascal was enthralled, and started performing experiments of his own, using water, wine, and other fluids. The result was *New experiments concerning the vacuum,* published in 1647. This work left the main question unanswered: why would mercury rise only 30 inches and water only 33 feet? The theories of the time tried to save a fragment of Aristotle's philosophy by declaring that nature's horror of the vacuum was "limited"; it could only destroy a finite amount of vacuum. Pascal had a different idea.

In the fall of 1648, acting on a hunch, Pascal sent his brother-in-law up a mountain with a mercury-filled tube. On

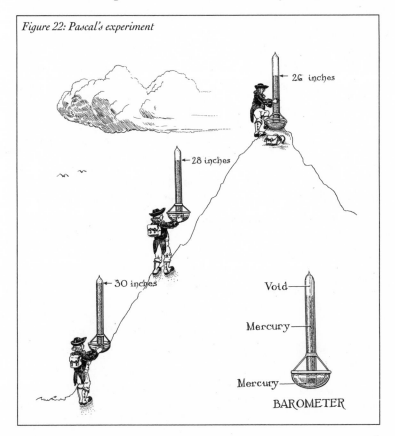

Figure 22: Pascal's experiment

← 26 inches

← 28 inches

← 30 inches

Void—

Mercury—

Mercury—

BAROMETER

top of the mountain, the mercury rose considerably less than 30 inches (Figure 22). Was nature somehow perturbed less by a vacuum on top of a mountain than by a vacuum in the valley?

To Pascal, this seemingly bizarre behavior proved that it wasn't an abhorrence of the vacuum that drove the mercury up the tube. It was the weight of the atmosphere pressing down on the mercury exposed in the pan that makes the fluid shoot up the column. The *atmospheric pressure* bearing down on a pan of liquid—be it mercury, water, or wine—will make the level inside the tube rise, just as gently squeezing the bottom of a toothpaste tube will make the contents squirt out the top. Since the atmosphere cannot push infinitely hard, it can only drive mercury about 30 inches up the tube—and at the top of the mountain, there is less atmosphere pushing down, so the air can't even push the mercury as high as 30 inches.

It is a subtle point: vacuums don't suck; the atmosphere pushes. But Pascal's simple experiment demolished Aristotle's assertion that nature abhors a vacuum. Pascal wrote, "But until now one could find no one who took this . . . view, that nature has no repugnance for the vacuum, that it makes no effort to avoid it, and that it admits vacuum without difficulty and without resistance." Aristotle was defeated, and scientists stopped fearing the void and began to study it.

It was also in zero and the infinite that Pascal, the devout Jansenist, sought to prove God's existence. He did it in a very profane way.

The Divine Wager

What is man in nature? Nothing in relation to the infinite, everything in relation to nothing, a mean between nothing and everything.

— BLAISE PASCAL, *PENSÉES*

Pascal was a mathematician as well as a scientist. In science Pascal investigated the vacuum—the nature of the void. In mathematics Pascal helped invent a whole new branch of the field: probability theory. When Pascal combined probability theory with zero and with infinity, he found God.

Probability theory was invented to help rich aristocrats win more money with their gambling. Pascal's theory was extremely successful, but his mathematical career was not to last. On November 23, 1654, Pascal had an intense spiritual experience. Perhaps it was the old Jansenist antiscience creed that was building up in him, but for whatever the reason, Pascal's newfound devotion led him to abandon mathematics and science altogether. (He made an exception for a brief time four years later, when he was unable to sleep owing to illness. He started doing mathematics and the pain eased. Pascal believed that this was a sign that God was not displeased with his studies.) He became a theologian—but he could not escape his profane past. Even when it came to arguing about God's existence, he kept coming back to those crass gambling Frenchmen. Pascal argued that it was best to believe in God, because it was a good bet. Literally.

Just as he analyzed the value—or expectation—of a gamble, Pascal analyzed the value of accepting Christ as savior. Thanks to the mathematics of zero and infinity, Pascal concluded that one should assume that God exists.

Before considering the wager itself, it is easy to analyze a slightly different game. Imagine that there are two envelopes, marked A and B. Before you are shown the envelopes, a flip of the coin determined which envelope has money in it. If the coin toss was a heads, A has a brand-new $100 bill inside. If the coin came up tails, B has the money—but this time, it's $1,000,000. Which envelope should you choose?

B, obviously! Its value is much greater. It is not difficult to show this using a tool from probability theory called an *expectation*, which is a measure of how much we expect each envelope to be worth.

Envelope A might or might not have a $100 bill in it; it has some value, because it might have money in it, but it isn't worth as much as $100, because you're not absolutely sure that it contains anything. In fact a mathematician would add up all of the possible contents of envelope A and then multiply by the probability of each outcome:

1/2 chance of winning $0	$1/2 \times \$0$	=	$0
1/2 chance of winning $100	$1/2 \times \$100$	=	$50
	Expectation	=	$50

The mathematician would conclude that the expected value of the envelope is $50. At the same time, the expected value of envelope B is:

1/2 chance of winning $0	$1/2 \times \$0$	=	$0
1/2 chance of winning $1,000,000	$1/2 \times \$1,000,000$	=	$500,000
	Expectation	=	$500,000

So the expected value of B is $500,000—10,000 times as much as the expected value of envelope A. Clearly, if you are offered a choice between the two envelopes, the smart thing to do is to choose B.

Pascal's wager is exactly like this game, except that it uses a different set of envelopes: Christian and atheist. (Actually, Pascal only analyzed the Christian case, but the atheist case is the logical extension.) For the sake of argument, imagine for the moment that there's a 50-50 chance that God exists. (Pas-

cal assumed that it would be the Christian God, of course.)
Now, choosing the Christian envelope is equivalent to choosing to be a devout Christian. If you happen to choose this
path, there are two possibilities. If you are a faithful Christian
and there is no God, you just fade into nothingness when you
die. But if there is a God, you go to heaven and live for eternity in bliss: infinity. So the expected value of being a Christian is:

1/2 chance of fading into nothing	$1/2 \times 0$	=	0
1/2 chance of going to heaven	$1/2 \times \infty$	=	∞
	Expectation	=	∞

After all, half of infinity is still infinity. Thus, the value of being a Christian is infinite. Now what happens if you are an
atheist? If you are correct—there is no God—you gain nothing from being right. After all, if there is no God, there is no
heaven. But if you are wrong and there is a God, you go to
hell for an eternity: negative infinity. So the expected value of
being an atheist is:

1/2 chance of fading into nothing	$1/2 \times 0$	=	0
1/2 chance of going to hell	$1/2 \times -\infty$	=	$-\infty$
	Expectation	=	$-\infty$

Negative infinity. The value is as bad as you can possibly get.
The wise person would clearly choose Christianity instead of
atheism.

But we made an assumption here—that there is a 50-50
chance that God exists. What happens if there is only a
1/1000 chance? The value of being a Christian would be:

999/1000 chance of fading into nothing	$999/1000 \times 0$ =	0
1/1000 chance of going to heaven	$1/1000 \times \infty$ =	∞
	Expectation	=	∞

It's still the same: infinite, and the value of being an atheist is
still negative infinity. It's still much better to be a Christian. If

the probability is 1/10,000 or 1/1,000,000 or one in a gazillion, it comes out the same. The only exception is zero.

If there is no chance that God exists, Pascal's wager—as it came to be known—makes no sense. The expected value of being a Christian would then be $0 \times \infty$, and that was gibberish. Nobody was willing to say that there was zero chance that God exists. No matter what your outlook, it is always better to believe in God, thanks to the magic of zero and infinity. Certainly Pascal knew which way to wager, even though he gave up mathematics to win his bet.

Chapter 5
Infinite Zeros and Infidel Mathematicians

[ZERO AND THE SCIENTIFIC REVOLUTION]

*With the introduction of . . . the infinitely small and infi-
nitely large, mathematics, usually so strictly ethical, fell
from grace. . . . The virgin state of absolute validity and
irrefutable proof of everything mathematical was gone
forever; the realm of controversy was inaugurated, and we
have reached the point where most people differentiate and
integrate not because they understand what they are doing but from pure faith, because up to now it has always
come out right.*

— FRIEDRICH ENGELS, *ANTI-DUHRING*

Zero and infinity had destroyed the Arisotelian philosophy; the void and the infinite cosmos had eliminated the
nutshell universe and the idea of nature's abhorrence of the
vacuum. The ancient wisdom was discarded, and scientists

began to divine the laws that governed the workings of nature. However, there was a problem with the scientific revolution: zero.

Deep within the scientific world's powerful new tool—calculus—was a paradox. The inventors of calculus, Isaac Newton and Gottfried Wilhelm Leibniz, created the most powerful mathematical method ever by dividing by zero and adding an infinite number of zeros together. Both acts were as illogical as adding 1 + 1 to get 3. Calculus, at its core, defied the logic of mathematics. Accepting it was a leap of faith. Scientists took that leap, for calculus is the language of nature. To understand that language completely, science had to conquer the infinite zeros.

The Infinite Zeros

When, after a thousand-year stupor, European thought shook off the effect of the sleeping powders so skilfully administered by the Christian Fathers, the problem of infinity was one of the first to be revived.

—TOBIAS DANZIG, *NUMBER: THE LANGUAGE OF SCIENCE*

Zeno's curse hung over mathematics for two millennia. Achilles seemed doomed to chase the tortoise forever, never catching up. Infinity lurked in Zeno's simple riddle. The Greeks were stumped by Achilles' infinite steps. They never considered adding infinite parts together even though Achilles' strides approach zero size; the Greeks could hardly add steps of zero size together without the concept of zero. However, once the West embraced zero, mathematicians began to tame the infinite and ended Achilles' race.

Even though Zeno's sequence has infinite parts, we can add all of the steps together and still stay within the realm of the finite: $1 + 1/2 + 1/4 + 1/8 + 1/16 + \ldots = 2$. The first person to do this sort of trick—adding infinite terms to get a finite result—was the fourteenth-century British logician Richard Suiseth. Suiseth took an infinite sequence of numbers: 1/2, 2/4, 3/8, 4/16, . . . , $n/2^n$, . . . , and added them all together, yielding two. After all, the numbers in the sequence get closer and closer to zero; naively, one would guess that this would ensure that the sum remains finite. Alas, the infinite is not quite that simple.

At about the same time Suiseth was writing, Nicholas Oresme, a French mathematician, tried his hand at adding together another infinite sequence of numbers—the so-called harmonic series:

$$1/2 + 1/3 + 1/4 + 1/5 + 1/6 + . . .$$

Like the Zeno sequence and Suiseth's sequence, all the terms get closer and closer to zero. However, when Oresme tried to sum the terms in the sequence, he realized that the sums got larger and larger and larger. Even though the individual terms go to zero, the sum goes off to infinity. Oresme showed this by clumping the terms together: $1/2 + (1/3 + 1/4) + (1/5 + 1/6 + 1/7 + 1/8) + ...$ The first group clearly equals 1/2; the second group is greater than $(1/4 + 1/4)$, or 1/2. The third group is greater than $(1/8 + 1/8 + 1/8 + 1/8)$, or 1/2. And so forth. You keep adding 1/2 after 1/2 after 1/2, and the sum gets bigger and bigger, and off to infinity. Even though the terms themselves go to zero, they don't approach zero fast enough. An infinite sum of numbers can be infinite, even if the numbers themselves approach zero. Yet this isn't the strangest aspect of infinite sums. Zero itself is not immune to the bizarre nature of infinity.

Consider the following series: $1 - 1 + 1 - 1 + 1 - 1 + 1 - 1 + 1 - . . . .$ It's not so hard to show that this series sums to zero. After all,

$$(1 - 1) + (1 - 1) + (1 - 1) + (1 - 1) + (1 - 1) + (1 - 1) + . . .$$

is the same thing as

$$0 + 0 + 0 + 0 + 0 + 0 + . . .$$

which clearly sums to zero. But beware! Group the series in a different way:

$$1 + (-1 + 1) + (-1 + 1) + (-1 + 1) + (-1 + 1) + (-1 + 1) + \ldots$$

is the same thing as

$$1 + 0 + 0 + 0 + 0 + 0 + \ldots$$

which clearly sums to one. The same infinite sum of zeros can equal 0 and 1 at the same time. An Italian priest, Father Guido Grandi, even used this series to prove that God could create the universe (1) out of nothing (0). In fact, the sequence can be set to equal anything at all. To make the sum equal 5, start with 5s and −5s instead of 1s and −1s, and we can show that $0 + 0 + 0 + 0 + \ldots$ equals 5.

Adding infinite things to each other can yield bizarre and contradictory results. Sometimes, when the terms go to zero, the sum is finite, a nice, normal number like 2 or 53. Other times the sum goes off to infinity. And an infinite sum of zeros can equal anything at all—and everything at the same time. Something very bizarre was going on; nobody knew quite how to handle the infinite.

Luckily the physical world made a little more sense than the mathematical one. Adding infinite things to each other seems to work out most of the time, so long as you are dealing with something in real life, like finding the volume of a barrel of wine. And 1612 was a banner year for wine.

Johannes Kepler—the man who figured out that planets move in ellipses—spent that year gazing into wine barrels, since he realized that the methods that vintners and coopers used to estimate the size of barrels were extremely crude. To help the wine merchants out, Kepler chopped up the barrels—in his mind—into an infinite number of infinitely tiny pieces, and then added them back together again to yield their volumes. This may seem a backward way of going about measuring barrels, but it was a brilliant idea.

To make the problem a bit simpler, let us consider a two-dimensional object rather than a three-dimensional one—a triangle. The triangle in Figure 23 has a height of 8 and a base of 8; since the area of a triangle is half the base times the height, the area is 32.

Now imagine trying to estimate the size of the triangle by inscribing little rectangles inside the triangle. For a first try, we get an area of 16, quite short of the actual value of 32. The second try is a bit better; with three rectangles, we get a value of 24. Closer, but still not there yet. The third try gives us 28—closer still. As you can see, making smaller and smaller

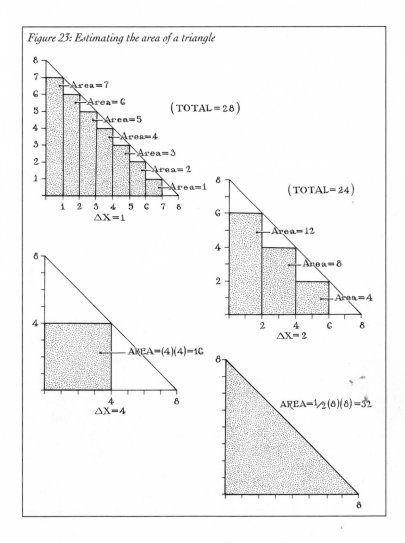

Figure 23: Estimating the area of a triangle

rectangles—whose widths, denoted by the symbol Δx, go to zero—makes the value closer and closer to 32, the true value for the area of the triangle. (The sum of these rectangles is equal to $\sum f(x) \Delta x$ where the Greek $\sum$ represents the sum over an appropriate range and $f(x)$ is the equation of the curve that the rectangles strike. In modern notation, as Δx goes to zero, we replace the $\sum$ with a new symbol, $\int$, and Δx with dx, turning the equation into $\int f(x)\ dx$, which is the integral.)

In one of Kepler's lesser-known works, *Volume-Measurement of Barrels*, he does this in three dimensions, slicing barrels into planes and summing the planes together. Kepler, at least, wasn't afraid of a glaring problem: as Δx goes to zero, the sum becomes equivalent to adding an infinite number of zeros together—a result that makes no sense. Kepler ignored the problem; though adding infinite zeros together was gibberish from a logical point of view, the answer it yielded was the right one.

Kepler was not the only prominent scientist who sliced objects infinitely thin. Galileo, too, pondered infinity and these infinitely small slices of area. These two ideas transcend our finite understanding, he wrote, "the former on account of their magnitude, the latter because of their smallness." Yet despite the deep mystery of the infinite zeros, Galileo sensed their power. "Imagine what they are when combined," he wondered. Galileo's student Bonaventura Cavalieri would provide part of the answer.

Instead of barrels, Cavalieri cut up geometric objects. To Cavalieri, every area, like that of the triangle, is made up of an infinite number of zero-width lines, and a volume is made up of an infinite number of zero-height planes. These *indivisible* lines and planes are like atoms of area and volume; they can't be divided any further. Just as Kepler measured the volumes of barrels with his thin slices, Cavalieri added up an infinite number of these indivisible zeros to figure out the area or the volume of a geometric object.

For geometers, Cavalieri's statement was troublesome indeed; adding infinite zero-area lines could not yield a two-dimensional triangle, nor could infinite zero-volume planes add up to a three-dimensional structure. It was the same problem: infinite zeros make no logical sense. However, Cavalieri's method always gave the right answer. Mathematicians ignored the logical and philosophical troubles with adding infinite zeros—especially since indivisibles or *infinitesimals*, as they came to be called, finally solved a long-standing puzzle: the problem of the tangent.

A tangent is a line that just kisses a curve. For any point along a smooth curve that flows through space, there is a line that just grazes the curve, touching at exactly one point. This is the tangent, and mathematicians realized that it is extremely important in studying motion. For instance, imagine swinging a ball on a string around your head. It's traveling in a circle. However, if you suddenly cut the string, the ball will fly off along that tangent line; in the same way, a baseball pitcher's arm travels in an arc as he throws, but as soon as he lets go, the ball flies off on the tangent (Figure 24). As another example, if you want to find out where a ball will come to rest at the bottom of a hill, you look for a point where the tangent line is horizontal. The steepness of the tangent line— its *slope*—has some important properties in physics: for instance, if you've got a curve that represents the position of, say, a bicycle, then the slope of the tangent line to that curve

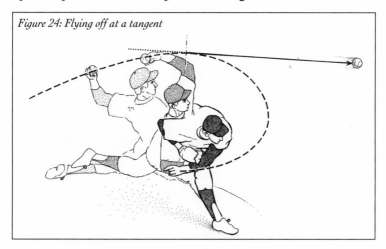

Figure 24: Flying off at a tangent

at any given point tells you how fast that bicycle is going when it reaches that spot.

For this reason, several seventeenth-century mathematicians—like Evangelista Torricelli, René Descartes, the Frenchman Pierre de Fermat (famous for his last theorem), and the Englishman Isaac Barrow—created different methods for calculating the tangent line to any given point on a curve. However, like Cavalieri, all of them came up against the infinitesimal.

To draw a tangent line at any given point, it's best to make a guess. Choose another point nearby and connect the two. The line you get isn't exactly the tangent line, but if the curve isn't too bumpy, the two lines will be pretty close. As you reduce the distance between the points, the guess gets closer to the tangent line (Figure 25). When your points are zero distance away from each other, your approximation is perfect: you have found the tangent. Of course, there's a problem.

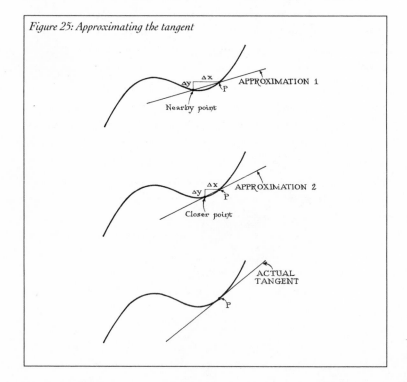

Figure 25: Approximating the tangent

The most important property of a line is its slope, and to measure this, mathematicians look at how high a line rises in a certain amount of distance. As an example, imagine you are driving east on a hill; for every mile east you drive, you gain half a mile in altitude. The slope of the hill is simply the height—half a mile—over the horizontal distance you have driven—one mile. Mathematicians say that the slope of the hill is 1/2. The same thing is true for lines; to measure the slope of a line, you look at how much the line rises (which mathematicians denote by the symbol Δy) in a given horizontal distance (which is denoted by Δx). The slope of the line is $\Delta y/\Delta x$.

When you try to calculate the slope of a tangent line, zero wrecks your approximation process. As your approximations of the tangent lines get better and better, the points on the curve you use to create the approximations get closer together. This means that the difference in height, Δy, goes to zero, as does the horizontal distance between the points, Δx. As your tangent approximations get better and better, $\Delta y/\Delta x$ approaches 0/0. Zero divided by zero can equal any number in the universe. Does the slope of the tangent line have any meaning?

Every time mathematicians tried to deal with the infinite or with zero, they encountered trouble with illogic. To figure out the volume of a barrel or the area under a parabola, mathematicians added infinite zeros together; to find out the tangent of a curve, they divided zero by itself. Zero and infinity made the simple acts of taking tangents and finding areas appear to be self-contradictory. These troubles would have ended as an interesting footnote but for one thing: these infinities and zeros are the key to understanding nature.

Zero and the Mystical Calculus

If we lift the veil and look underneath . . . we shall dis-
cover much emptiness, darkness, and confusion; nay, if I
mistake not, direct impossibilities and contradictions. . . .
They are neither finite quantities, nor quantities infi-
nitely small, nor yet nothing. May we not call them the
ghosts of departed quantities?

— BISHOP BERKELEY, *THE ANALYST*

The tangent problem and the area problem both ran afoul of the same difficulties with infinities and zeros. It's no wonder, because the tangent problem and the area problem are actually the same thing. They are both aspects of calculus, a scientific tool far more powerful than anything ever seen before. The telescope, for instance, had given scientists the ability to find moons and stars that had never been observed before. Calculus, on the other hand, gave scientists a way to express the laws that govern the motion of the celestial bodies—and laws that would eventually tell scientists how those moons and stars had formed. Calculus was the very language of nature, yet its very fabric was infused with zeros and infinities that threatened to destroy the new tool.

The first discoverer of calculus nearly died before he ever took a breath. Born prematurely on Christmas Day in 1642, Isaac Newton squirmed into the world, so small that he was able to fit into a quart pot. His father, a farmer, had died two months earlier.

Despite a traumatic childhood* and a mother who wanted

---

* When Newton was three, his mother remarried and moved. Newton didn't accompany his mother and stepfather. As a result, he had little contact with his parents after that, unless you count the time he threatened to come over and burn their house down with them inside.

him to become a farmer, Newton enrolled in Cambridge in the 1660s—and flourished. Within a few years he developed a systematic method of solving the tangent problem; he could figure out the tangent to any smooth curve at any point. This process, the first half of calculus, is now known as differentiation; however, Newton's method of differentiation doesn't look very much like the one we use today.

Newton's style of differentiation was based upon *fluxions*— the flows—of mathematical expressions that he called *fluents*. As an example of Newton's fluxions, take the equation

$$y = x^2 + x + 1$$

In this equation, the fluents are y and x; Newton supposed that y and x are changing, or flowing, as time progresses. Their rates of change—their fluxions—are denoted by $\dot{y}$ and $\dot{x}$ respectively.

Newton's method of differentiation was based on a notational trick: he let the fluxions change, but he only let them change infinitesimally. Essentially, he gave them no time to flow. In Newton's notation, y would change in that instant to $(y + o\dot{y})$ while x changes to $(x + o\dot{x})$. (The letter o represented the amount of time that had passed; it was almost a zero, but not quite, as we shall see.) The equation then becomes

$$(y + o\dot{y}) = (x + o\dot{x})^2 + (x + o\dot{x}) + 1$$

Multiplying out the $(x + o\dot{x})^2$ term gives us

$$y + o\dot{y} = x^2 + 2x(o\dot{x}) + (o\dot{x})^2 + x + o\dot{x} + 1$$

Rearranging the terms yields

$$y + o\dot{y} = (x^2 + x + 1) + 2x(o\dot{x}) + 1(o\dot{x}) + (o\dot{x})^2$$

Since $y = x^2 + x + 1$, we can subtract y from the left side of the equation and $x^2 + x + 1$ from the right side of the equation and leave the system unchanged. That leaves us with

$$o\dot{y} = 2x(o\dot{x}) + 1(o\dot{x}) + (o\dot{x})^2$$

Now comes the dirty trick. Newton declared that since $o\dot{x}$ was really, really small, $(o\dot{x})^2$ was even smaller: it vanished. In essence, it was zero, and could be ignored. That gives us

$$o\dot{y} = 2x(o\dot{x}) + 1(o\dot{x})$$

which means that $o\dot{y}/o\dot{x} = 2x + 1$, which is the slope of the tangent line at any point x on the curve (Figure 26). The infinitesimal time period o drops right out of the equation, $o\dot{y}/o\dot{x}$ becomes $\dot{y}/\dot{x}$, and o need never be thought of again.

The method gave the right answer, but Newton's vanishing act was very troubling. If, as Newton insisted, $(o\dot{x})^2$ and $(o\dot{x})^3$ and higher powers of $o\dot{x}$ were equal to zero, then $o\dot{x}$ it-

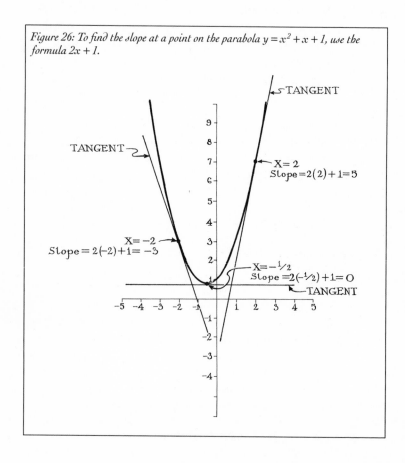

Figure 26: To find the slope at a point on the parabola $y = x^2 + x + 1$, use the formula $2x + 1$.

self must be equal to zero.* On the other hand, if $o\dot{x}$ was zero, then dividing by $o\dot{x}$ as we do toward the end is the same thing as dividing by zero—as is the very last step of getting rid of the o in the top and bottom of the $o\dot{y}/o\dot{x}$ expression. Division by zero is forbidden by the logic of mathematics.

Newton's method of fluxions was very dubious. It relied upon an illegal mathematical operation, but it had one huge advantage. It worked. The method of fluxions not only solved the tangent problem, it also solved the area problem. Finding the area under a curve (or a line, which is a type of curve)—an operation we now call integration—is nothing more than the reverse of differentiation. Just as differentiating the curve $y = x^2 + x + 1$ gives you an equation for the slope of the tangent—$y = 2x + 1$—integrating the curve $y = 2x + 1$ gives you a formula for the area under the curve. This formula is $y = x^2 + x + 1$; the area underneath the curve between the boundaries $x = a$ and $x = b$ is simply $(b^2 + b + 1) - (a^2 + a + 1)$ (Figure 27). (Technically, the formula is $y = x^2 + x + c$, where c is any constant you choose. The process of differentiation destroys information, so the process of integration doesn't give you exactly the answer you are looking for unless you add another bit of information.)

Calculus is the combination of these two tools, differentiation and integration, in one package. Though Newton broke some very important mathematical rules by toying with the powers of zero and infinity, calculus was so powerful that no mathematician could reject it.

Nature speaks in equations. It is an odd coincidence. The rules of mathematics were built around counting sheep and surveying property, yet these very rules govern the way the universe works. Natural laws are described with equations, and equations, in a sense, are simply tools where you plug in

---

* If you multiply two numbers together and get zero, then one or the other must equal zero. (In mathematical terms, if $ab = 0$, then $a = 0$ or $b = 0$.) This means that if $a^2 = 0$, then $aa = 0$, thus $a = 0$.

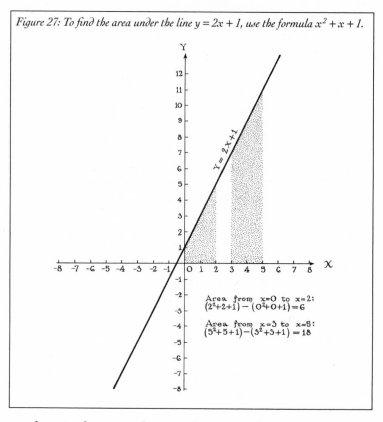

Figure 27: To find the area under the line $y = 2x + 1$, use the formula $x^2 + x + 1$.

Area from x=0 to x=2:
$(2^2+2+1) - (0^2+0+1) = 6$

Area from x=3 to x=5:
$(5^2+5+1) - (3^2+3+1) = 18$

numbers and get another number out. The ancients knew a few of these equation-laws, like the law of the lever, but with the beginning of the scientific revolution these equation-laws sprang up everywhere. Kepler's third law described the time it takes for planets to complete an orbit: $r^3/t^2 = k$ for time t, distance r, and a constant k. In 1662, Robert Boyle showed that if you take a sealed container with a gas in it, squishing the container would increase the pressure inside: pressure p times volume v was always a constant — $pv = k$ for a constant k. In 1676, Robert Hooke figured out that the force exerted by a spring, f, was a negative constant, $-k$, multiplied by the distance, x, that you've stretched it: $f = -kx$. These early equation-laws were extremely good at expressing simple relationships, but equations have limitations — their constancy, which prevented them from being universal laws.

As an example, let's take the famous equation we all learned in high school: rate times time equals distance. It shows how far you get, x miles, when you run with a certain velocity, v miles per hour, for a time, t hours: $vt = x$; after all, miles per hour times hours equals miles. This equation is very useful when you are calculating how long it will take to get from New York to Chicago on a train that moves exactly 120 miles an hour. But how many things really move at a constant rate like a train in a math problem? Drop a ball, and it moves faster and faster; in this case, $x = vt$ is quite simply wrong. For the case of a dropped ball, $x = gt^2/2$, where g is the acceleration due to gravity. On the other hand, if you've got an increasing force on the ball, x might equal something like $t^3/3$. Rate times time equals distance is not a universal law; it doesn't apply under all conditions.

Calculus allowed Newton to combine all these equations into one grand set of laws—laws that applied in all cases, under all conditions. For the first time, science could see the universal laws that underlie all of these little half laws. Even though mathematicians knew that calculus was deeply flawed—thanks to the mathematics of zero and infinity—they quickly embraced the new mathematical tools. For the truth is, nature doesn't speak in ordinary equations. It speaks in *differential equations*, and calculus is the tool that you need to pose and solve these differential equations.

Differential equations are not like the everyday equations that we are all familiar with. An everyday equation is like a machine; you feed numbers into the machine and out pops another number. A differential equation is also like a machine, but this time you feed equations into the machine and out pop new equations. Plug in an equation that describes the conditions of the problem (is the ball moving at a constant rate, or is a force acting on the ball?) and out pops the equation that encodes the answer that you seek (the ball moves in a straight line or in a parabola). One differential equation governs all of the uncountable numbers of equation-laws. And unlike the little equation-laws that sometimes hold and sometimes don't, the differential equation is always true. It is a universal law. It is a glimpse at the machinery of nature.

Newton's calculus—his method of fluxions—did just this by tying together concepts like position, velocity, and acceleration. When Newton denoted position with the variable x, he realized that velocity is simply the fluxion—what modern mathematicians call the derivative—of x: $\dot{x}$. And acceleration is nothing more than the derivative of velocity, $\ddot{x}$. Going from position to velocity to acceleration and back again is as simple as differentiating (adding another dot) or integrating (removing a dot). With that notation in hand, Newton was able to create a simple differential equation that describes the motion of all objects in the universe: $F = m\ddot{x}$, where F is the force on an object and m is its mass. (Actually, this is not quite a universal law, as the equation only holds when the mass of an object is constant. The more general version of Newton's law is $F = \dot{p}$, where p is an object's momentum. Of course, Newton's equations were eventually refined further by Einstein.) If you've got an equation that tells you about the force that is being applied on an object, the differential equation reveals exactly how the object moves. For instance, if you have a ball in free fall, it moves in a parabola, while a frictionless spring wobbles back and forth forever, and a spring with friction slowly comes to rest (Figure 28). As different as these outcomes seem, they are all governed by the same differential equation.

Likewise, if you know the way an object moves—whether it be a toy ball or a giant planet—the differential equation can tell you what kind of force is being applied. (Newton's triumph was taking the equation that described the force of gravity and figuring out the shapes of the planets' orbits. People had suspected that the force was proportional to $1/r^2$, and when ellipses popped right out of Newton's differential equations, people began to believe that Newton was correct.) Despite the power of calculus, the key problem remained. Newton's work was based on a very shaky foundation—dividing zero by itself. His rival's work had the same flaw.

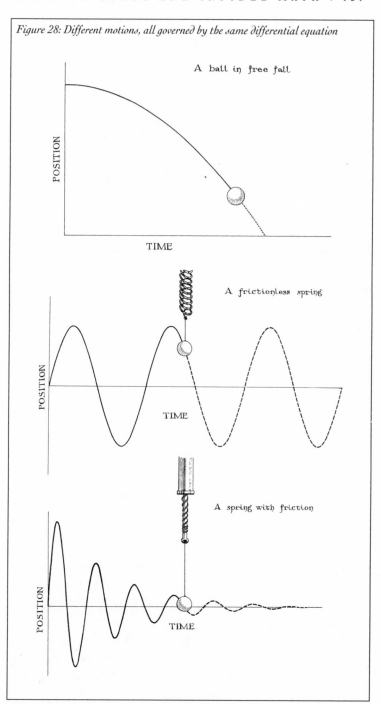

Figure 28: Different motions, all governed by the same differential equation

A ball in free fall

POSITION

TIME

A frictionless spring

POSITION

TIME

A spring with friction

POSITION

TIME

In 1673 an esteemed German lawyer and philosopher visited London. His name was Gottfried Wilhelm Leibniz. He and Newton would tear the scientific world asunder, though neither would solve the problem of the zeros that suffused calculus.

Nobody knows whether the thirty-three-year-old Leibniz encountered Newton's unpublished work during his trip to England. But between 1673 and 1676, when Leibniz next visited London, he, too, had developed calculus, although in a slightly different form.

Looking back, it appears that Leibniz formulated his version independently of Newton, though the matter is still being debated. The two had a correspondence in the 1670s, making it very difficult to establish how they influenced each other. However, though the two theories came up with the same answers, their notations — and their philosophies — were very different.

Newton disliked infinitesimals, the little os in his fluxion equations that sometimes acted like zeros and sometimes like nonzero numbers. In a sense these infinitesimals were infinitely small, smaller than any positive number you could name, yet still somehow greater than zero. To the mathematicians of the time, this was a ridiculous concept. Newton was embarrassed by the infinitesimals in his equations, and he swept them under the rug. The os in his calculations were only intermediaries, crutches that vanished miraculously by the end of the computation. On the other hand, Leibniz reveled in the infinitesimal. Where Newton wrote $o\dot{x}$, Leibniz wrote dx — an infinitesimally tiny little piece of x. These infinitesimals survived unchanged throughout Leibniz's calculations; indeed the derivative of y with respect to x was not the infinitesimal-free ratio of fluxions $\dot{y}/\dot{x}$, but the ratio of infinitesimals dy/dx.

With Leibniz's calculus, these dys and dxs can be manipulated just like ordinary numbers, which is why modern mathematicians and physicists usually use Leibniz's notation rather than Newton's. Leibniz's calculus had the same power as Newton's, and thanks to its notation, even a bit more. Nevertheless, underneath all the mathematics, Leibniz's differentials still had the same forbidden 0/0 nature that plagued Newton's fluxions. As long as this flaw remained, calculus would be based upon faith rather than logic. (In fact, faith was very much on Leibniz's mind when he derived new mathematics, such as the *binary numbers*. Any number can be written as a string of zeros and ones; to Leibniz, this was the *creation ex nihilo*, the creation of the universe out of nothing more than God/1 and void/0. Leibniz even tried to get the Jesuits to use this knowledge to convert the Chinese to Christianity.)

It would be many years before mathematicians began to free calculus from its mystical underpinnings, for the mathematical world was busy fighting over who invented calculus.

There is little doubt that Newton came up with the idea first—in the 1660s—but he did not publish his work for 20 years. Newton was a magician, theologian, and alchemist as well as a scientist (for instance, he used biblical texts to conclude that the second coming of Christ would occur around 1948) and many of his views were heretical. As a result, he was secretive and reluctant to reveal his work. In the meantime, while Newton sat upon his discovery, Leibniz developed his own calculus. The two promptly accused each other of plagiarism, and the English mathematical community, which backed Newton, pulled away from the Continental mathematicians, who supported Leibniz. As a result, the English stuck to Newton's fluxion notation rather than adopting Leibniz's superior differential notation—cutting off their noses to spite their faces. English mathematicians fell far behind their Continental counterparts when it came to developing calculus.

A Frenchman, not an Englishman, would be remembered for taking the first nibble at the mysterious zeros and infinities that suffused calculus; mathematicians learn of l'Hôpital's rule when they first learn about calculus. Oddly enough, it was not l'Hôpital who came up with the rule that bears his name.

Born in 1661, Guillaume-François-Antoine de l'Hôpital was a marquis—and was thus very wealthy. He had an early interest in mathematics, and though he spent some time in the army, becoming a cavalry captain, he soon turned back to his true love of math.

L'Hôpital bought himself the best teacher that money could buy: Johann Bernoulli, a Swiss mathematician and one of the early masters of Leibniz's calculus of infinitesimals. In 1692, Bernoulli taught l'Hôpital calculus. L'Hôpital was so enthralled by the new mathematics that he persuaded Bernoulli to send him all Bernoulli's new mathematical discoveries for the marquis to use as he desired, in return for cash. The result was a textbook. In 1696, l'Hôpital's *Analyse des infiniment petits* became the first textbook on calculus and introduced much of Europe to the Leibnizian version. Not only did l'Hôpital explain the fundaments of calculus in his textbook, he also included some exciting new results. The most famous is known as l'Hôpital's rule.

L'Hôpital's rule took the first crack at the troubling 0/0 expressions that were popping up throughout calculus. The rule provided a way to figure out the true value of a mathematical function that goes to 0/0 at a point. L'Hôpital's rule states that the value of the fraction was equal to the derivative of the top expression divided by the derivative of the bottom expression. For instance, consider the expression $x/(\sin x)$ when $x = 0$; $x = 0$, as does $\sin x$, so the expression is equal to 0/0. Using L'Hopital's rule, we see that the expression goes to $1/(\cos x)$, as 1 is the derivative of x and $\cos x$ is the derivative of $\sin x$. $\cos x = 1$ when $x = 0$, so the whole expression equals $1/1 = 1$. Clever manipulations could also bring l'Hopital's rule to resolve other odd expressions: ∞/∞, 0^0, 0^∞, and ∞^0.

All of these expressions, but especially 0/0, could take on any value you desire them to have, depending on the functions you put in the numerator and denominator. This is why 0/0 is dubbed *indeterminate*. It was no longer a complete mystery; mathematicians could extract some information about 0/0 if they approached it very carefully. Zero was no longer an enemy to be avoided; it was an enigma to be studied.

Soon after l'Hôpital's death in 1704, Bernoulli started implying that l'Hôpital had stolen his work. At the time the mathematical community rejected Bernoulli's claims; not only had l'Hôpital proved himself an able mathematician, but Johann Bernoulli had a tarnished reputation. He had previously tried to claim credit for another mathematician's proof. (The other mathematician happened to be his brother, Jakob.) In this case, though, Johann Bernoulli's claim was justified. His correspondence with l'Hôpital backs his story. Alas for Bernoulli, the name for l'Hôpital's rule stuck.

L'Hôpital's rule was extremely important for resolving some of the difficulties with 0/0, but the underlying problem remained. Newton's and Leibniz's calculus depends upon dividing by zero—and on numbers that miraculously disappear when you square them. L'Hôpital's rule examines 0/0 with tools that were built upon 0/0 to begin with. It is a circular argument. And as physicists and mathematicians all over the world were beginning to use calculus to explain nature, and the concept of absolute space to explain motion, cries of protest emanated from the church.

In 1734, seven years after Newton's death, an Irish bishop, George Berkeley, wrote a book entitled *The Analyst, Or a Discourse Addressed to an Infidel Mathematician*. (The mathematician in question was most likely Edmund Halley, always a supporter of Newton.) In *The Analyst*, Berkeley pounced on Newton's (and Leibniz's) dirty tricks with zeros.

Calling infinitesimals "ghosts of departed quantities," Berkeley showed how making these infinitesimals disappear with impunity can lead to a contradiction. He concluded that "he who can digest a second or third fluxion, a second or third difference, need not, methinks, be squeamish about any point in divinity."

Though mathematicians of the day sniped at Berkeley's logic, the good bishop was entirely correct. In those days calculus was very different from other realms of mathematics. Every theorem in geometry had been rigorously proved; by taking a few rules from Euclid and proceeding, very carefully, step by step, a mathematician could show how a triangle's angles sum to 180 degrees, or any other geometric fact. On the other hand, calculus was based on faith.

Nobody could explain how those infinitesimals disappeared when squared; they just accepted the fact because making them vanish at the right time gave the correct answer. Nobody worried about dividing by zero when conveniently ignoring the rules of mathematics explained everything from the fall of an apple to the orbits of the planets in the sky. Though it gave the right answer, using calculus was as much an act of faith as declaring a belief in God.

The End of Mysticism

A quantity is something or nothing; if it is something, it has not yet vanished; if it is nothing, it has literally vanished. The supposition that there is an intermediate state between these two is a chimera.

—JEAN LE ROND D'ALEMBERT

In the shadow of the French Revolution, the mystical was driven out of calculus.

Despite calculus's shaky foundations, by the end of the eighteenth century, mathematicians all over Europe were having stunning successes with the new tool. Colin Maclaurin and Brook Taylor, perhaps the best British mathematicians in the era of isolation from the Continent, discovered how to use calculus to rewrite functions in a totally different form. For instance, after using some tricks in calculus, mathematicians realized that the function $1/(1 - x)$ can be written as

$$1 + x + x^2 + x^3 + x^4 + x^5 + \ldots$$

Though the two expressions look dramatically different, they are (with some caveats) exactly the same.

Those caveats, which stem from the properties of zero and infinity, can become very important, however. The Swiss mathematician Leonhard Euler, inspired by calculus's easy manipulation of zeros and infinities, used similar reasoning as Taylor and Maclaurin and "proved" that the sum

$$\ldots 1/x^3 + 1/x^2 + 1/x + 1 + x + x^2 + x^3 \ldots$$

equals zero. (To convince yourself that something fishy is going on, plug in the number 1 for x and see what happens.) Euler was an excellent mathematician — in fact, he was one of the most prolific and influential in history — but in this case the careless manipulation of zero and infinity led him astray.

It was a foundling who finally tamed the zeros and infinities in calculus and rid mathematics of its mysticism. In 1717 an infant was found on the steps of the church of Saint Jean Baptiste le Rond in Paris. In memory of that occasion, the child was named Jean Le Rond, and he eventually took the surname d'Alembert. Though he was raised by an impoverished working-class couple — his foster father was a glazier — it turns out that his birth father was a general and his mother was an aristocrat.

D'Alembert is best known for his collaboration on the famed *Encyclopédie* of human knowledge — a 20-year effort with coauthor Denis Diderot. But d'Alembert was more than an encyclopedist. It was d'Alembert who realized that it was important to consider the journey as well as the destination. He was the one who hatched the idea of *limit* and solved calculus's problems with zeros.

Once again, let us consider the story of Achilles and the tortoise, which is an infinite sum of steps that get closer and closer to zero. Manipulating an infinite sum — whether it is in the Achilles problem or in finding the area underneath a curve or finding an alternate form for a mathematical function — caused mathematicians to come up with contradictory results.

D'Alembert realized that the Achilles problem vanishes if

you consider the *limit* of the race. In our example on page 41, at every step the tortoise and Achilles get closer and closer to the two-foot mark. No step takes them farther away or even keeps them at the same distance; each moment brings them closer to that mark. Thus, the limit of that race—its ultimate destination—is at the two-foot mark. This is where Achilles passes the tortoise.

But how do you prove that two feet is actually the limit of the race? I ask you to challenge me. Give me a tiny distance, no matter how small, and I will tell you when both Achilles and the tortoise are less than that tiny distance away from the limit.

As an example, let's say that you challenge me with a distance of one-thousandth of a foot. Well, a few calculations later, I would tell you that after the 11th step, Achilles is 977 millionths of a foot away from the two-foot mark, while the tortoise is half that distance away; I have met your challenge with 23 millionths of a foot to spare. What if you challenged me with a distance of one-billionth of a foot? After 31 steps, Achilles is 931 trillionths of a foot away from the target—69 trillionths closer than you needed—while the tortoise, again, is half that distance away. No matter how you challenge me, I can meet that challenge by telling you a time when Achilles is closer to the mark than you require. This shows that, indeed, Achilles is getting arbitrarily close to the two-foot mark as the race progresses: two feet is the limit of the race.

Now, instead of thinking of the race as a sum of infinite parts, think of it as a limit of finite sub-races. For instance, in the first race Achilles runs to the one-foot mark. Achilles has run

$$1$$

1 foot in all. In the next race Achilles does the first two parts—first running 1 foot, and then a half foot. In total, Achilles has run

$$1 + 1/2$$

1.5 feet in all. The third race takes him as far as

$$1 + 1/2 + 1/4$$

1.75 feet, all told. Each of these sub-races is finite and well-defined; we never encounter an infinity.

What d'Alembert did informally—and what the Frenchman Augustin Cauchy, the Czech Bernhard Bolzano, and the German Karl Weierstrass would later formalize—was to rewrite the infinite sum

$$1 + 1/2 + 1/4 + 1/8 + \ldots + 1/2^n + \ldots$$

as the expression

limit (as n goes to ∞) of $1 + 1/2 + 1/4 + 1/8 + \ldots + 1/2^n$

It's a very subtle change in notation, but it makes all the difference in the world.

When you have an infinity in an expression, or when you divide by zero, all the mathematical operations—even those as simple as addition, subtraction, multiplication, and division—go out the window. Nothing makes sense any longer. So when you deal with an infinite number of terms in a series, even the + sign doesn't seem so straightforward. That is why the infinite sum of +1 and −1 we saw at the beginning of the chapter seems to equal 0 and 1 at the same time.

However, by putting this limit sign in front of a series, you separate the process from the goal. In this way you avoid manipulating infinities and zeros. Just as Achilles' sub-races are each finite, each *partial sum* in a limit is finite. You can add them, divide them, square them; you can do whatever you want. The rules of mathematics still work, since everything is finite. Then, after all your manipulations are complete, you take the limit: you extrapolate and figure out where the expression is headed.

Sometimes that limit doesn't exist. For instance, the infinite sum of +1 and −1 does not have a limit. The value of the partial sums flips back and forth between 1 and 0; it's not really heading to a predictable destination. But with Achilles' race, the partial sums go from 1 to 1.5 to 1.75 to 1.875 to 1.9375 and so forth; they get closer and closer to two. The sums have a destination—a limit.

The same thing goes for taking the derivative. Instead of dividing by zero as Newton and Leibniz did, modern mathematicians divide by a number that they let approach zero. They do the division—perfectly legally, since there are no zeros—then they take the limit. The dirty tricks of making squared infinitesimals disappear and then dividing by zero to get a derivative were no longer necessary (see appendix C).

This logic may seem like splitting hairs, like an argument as mystical as Newton's "ghosts," but in reality it's not. It satisfies the mathematician's strict requirement of logical rigor. There is a very firm, consistent basis for the concept of limits. Indeed, you can even dispense with the "I challenge you" argument entirely, as there are other ways of defining a limit, such as calling it the convergence of two numbers, the *lim sup* and *lim inf.* (I have a wonderful proof of this, but alas, this book is too small to contain it.) Since limits are logically airtight, by defining a derivative in terms of limits, it becomes airtight as well—and puts calculus on a solid foundation.

No longer was it necessary to divide by zeros. Mysticism vanished from the realm of mathematics and logic ruled once more. The peace lasted until the Reign of Terror.

Chapter **6**
Infinity's Twin

[THE INFINITE NATURE OF ZERO]

God made integers; all else is the work of man.

— LEOPOLD KRONECKER

Zero and infinity always looked suspiciously alike. Multiply zero by anything and you get zero. Multiply infinity by anything and you get infinity. Dividing a number by zero yields infinity; dividing a number by infinity yields zero. Adding zero to a number leaves the number unchanged. Adding a number to infinity leaves infinity unchanged.

These similarities were obvious since the Renaissance, but mathematicians had to wait until the end of the French Revolution before they finally unraveled zero's big secret.

Zero and infinity are two sides of the same coin—equal and opposite, yin and yang, equally powerful adversaries at either end of the realm of numbers. The troublesome nature of zero lies with the strange powers of the infinite, and it is possible to understand the infinite by studying zero. To learn this, mathematicians had to venture into the world of the imaginary, a bizarre world where circles are lines, lines are circles, and infinity and zero sit on opposite poles.

The Imaginary

...a fine and wonderful refuge of the divine spirit —
almost an amphibian between being and non-being.

—GOTTFRIED WILHELM LEIBNIZ

Zero is not the only number that was rejected by mathematicians for centuries. Just as zero suffered from Greek prejudice, other numbers were ignored as well, numbers that made no geometric sense. One of these numbers, i, held the key to zero's strange properties.

Algebra presented another way of looking at numbers, entirely divorced from the Greek geometric ideas. Instead of trying to measure the area inside a parabola as the Greeks did, early algebraists sought to find the solutions to equations that encode relationships between different numbers. For instance, the simple equation $4x - 12 = 0$ describes how an unknown number x is related to 4, 12, and 0. The task of the algebra student is to figure out what number x is. In this case x is 3. Substitute 3 for x in the above equation and you will quickly see that the equation is satisfied; 3 is a solution for the equation $4x - 12 = 0$. In other words, 3 is a *zero* or a *root* of the expression $4x - 12$.

When you start stringing symbols together to get equations, you can wind up with something unexpected. For instance, take the above equation and change the − sign into a + sign. This leaves us with a very innocent-looking equation, $4x + 12 = 0$, but the solution to that equation is now −3, a negative number.

Just as Indian mathematicians accepted zero while Europeans rejected it for centuries, the East embraced negative numbers while the West tried to ignore them. As late as the seventeenth century, Descartes refused to accept negative numbers as roots of equations. He called them "false roots," which explains why he never extended his coordinate system to the negative numbers. Descartes was a late holdover, a victim of his success in marrying algebra to geometry. Negative numbers had long been useful to algebraists—even Western algebraists. Negative numbers came up all the time in solving equations, such as quadratic equations.

A *linear* equation like $4x - 12 = 0$ is extremely simple to solve, and such problems didn't entertain algebraists for very long. So they soon turned to more difficult problems: quadratic equations—equations that begin with an x^2 term, like $x^2 - 1 = 0$. Quadratic equations are more complicated than regular equations; for one thing, they can have two different roots. For instance, $x^2 - 1 = 0$ has two solutions: 1 and -1. (Substitute -1 or 1 for x in the equation and you'll see what happens.) Either one of those solutions works; as it turns out, the expression $x^2 - 1$ *splits* into $(x - 1)(x + 1)$, making it easy to see that if x is 1 or -1, the expression goes to zero.

Though quadratic equations are more complicated than linear equations, there is a simple way to figure out what the roots of a quadratic equation are. It's the famous quadratic formula, which is the crowning achievement of high-school algebra class. The formula for finding the roots of a quadratic equation $ax^2 + bx + c = 0$ is: $x = \dfrac{-b \pm \sqrt{b^2 - 4ac}}{2a}$. The + sign gives us one root, while the − sign gives us the other. The quadratic formula has been known for centuries; the ninth-century mathematician al-Khowarizmi knew how to solve almost every quadratic equation, though he didn't seem to consider negative numbers as roots. Not long after that, algebraists learned to accept negative numbers as valid solutions to equations. Imaginary numbers, though, were a little different.

Imaginary numbers never appeared in linear equations, but they began to crop up in quadratic ones. Consider the equation $x^2 + 1 = 0$. No number seems to solve the equation; plugging in –1, 3, –750, 235.23, or any other positive or negative number you could think of doesn't yield the correct answer. The expression simply will not split. Worse yet, when you try to apply the quadratic equation, you get two silly-sounding answers:

$$+\sqrt{-1} \; and \; -\sqrt{-1}$$

These expressions don't seem to make any sense. The Indian mathematician Bhaskara wrote in the twelfth century that "there is no square root of a negative number, for a negative number is not a square." What Bhaskara and others realized was that when you square a positive number, you get a positive number back; 2 times 2 equals 4, for instance. When you square a negative number, you still get a positive number: –2 times –2 also equals 4. When you square zero, you get zero. Positive numbers, negative numbers, and zero all give you nonnegative squares, and those three possibilities cover the whole number line. This means that there is no number on the number line that gives you a negative number when you square it. The square root of a negative number seemed like a ridiculous concept.

Descartes thought that these numbers were even worse than negative numbers; he came up with a scornful name for the square roots of negatives: *imaginary numbers.* The name stuck, and eventually, the symbol for the square root of –1 became *i.*

Algebraists loved *i.* Almost everyone else hated it. It was wonderful for solving polynomials—expressions like $x^3 + 3x + 1$ that have x raised to various powers. In fact, once you allow i into the realm of numbers, every polynomial becomes solvable: $x^2 + 1$ suddenly splits into $(x - i)(x + i)$—the roots of the equation are $+i$ and $-i$. Cubic expressions like $x^3 - x^2 + x - 1$ split three ways, such as $(x - 1)(x + i)(x - i)$. Quartic expressions—ones with a leading x^4 term—always split into four terms, and quintics—ones with a leading x^5

term—split five ways. All polynomials of degree n—those that have a leading term of x^n—split into n distinct terms. This is the *fundamental theorem of algebra*.

As early as the sixteenth century, mathematicians were using numbers with i included—the so-called *complex numbers*—to solve cubic and quartic polynomials. And while many mathematicians saw the complex numbers as a convenient fiction, others saw God.

Leibniz thought that i was a bizarre mix between existence and nonexistence, something like a cross between 1 (God) and 0 (Void) in his binary scheme. Leibniz likened i to the Holy Spirit: both have an ethereal and barely substantial existence. But even Leibniz didn't realize that i would finally reveal the relationship between zero and infinity. It would take two important developments in mathematics before the true link was uncovered.

Point and Counterpoint

One will then see the simplicity with which these concepts lead to properties already known and to an infinity of others which ordinary geometry does not seem to touch easily.

— JEAN-VICTOR PONCELET

The first development—projective geometry—was born in the turmoil of war. In the 1700s, France, England, Austria, Prussia, Spain, the Netherlands, and other countries were vying for power. Alliances formed and broke over and over again; new territorial disputes erupted over colonies, and countries struggled to dominate trade to and from the New World. France, England, and other countries skirmished throughout the first half of the eighteenth century, and roughly a quarter century after Newton died, a full-scale war erupted. France, Austria, Spain, and Russia fought England and Prussia for nine years.

In 1763 the French capitulated and the Seven Years' War was over. (Two years of fighting occurred before war was officially declared.) The victory made England the preeminent power in the world, but it came at a great cost. Both France and England were exhausted and in debt—and they would both suffer the consequences: revolutions. A little more than a decade after the end of the Seven Years' War, the American Revolution began; the revolt would strip England of its richest colony. In 1789, just as George Washington was sworn into office in the newly founded United States, the French Revolution began. Four years later the revolutionaries removed the French king's head.

A mathematician, Gaspard Monge, signed the revolutionary government's record of the king's execution. Monge was a consummate geometer, specializing in three-dimensional geometry. He was responsible for the way architects and engineers draw buildings and machines: they project the design onto a vertical plane and a horizontal plane, preserving all the information needed to reconstruct the object. Monge's work was so important to the military that much of it was made into a state secret by the revolutionary government and by the Napoleonic government that succeeded it soon afterward.

Jean-Victor Poncelet was a student of Monge's who learned about three-dimensional geometry as he trained to become an engineer for Napoleon's army. Unluckily for Poncelet, he entered the army just as Napoleon set off for Moscow in 1812.

While retreating from Moscow, Napoleon's army was whittled down to almost nothing by a harsh winter and an equally harsh Russian army. At the battle of Krasnoy, Poncelet was left for dead on the battlefield. Still alive, he was captured by the Russians. Moldering in a Russian prison, Poncelet founded a new discipline: projective geometry.

Poncelet's mathematics was the culmination of the work begun by the artists and architects of the fifteenth century, like Filippo Brunelleschi and Leonardo da Vinci, who discovered how to draw realistically—in perspective. When "parallel" lines converge at the vanishing point in a painting, observers are tricked into believing that the lines never meet. Squares on the floor become trapezoids in a painting; everything gets gently distorted, but it looks perfectly natural to the viewer. This is the property of an infinitely distant point—a zero at infinity.

Johannes Kepler, the man who discovered that planets travel in ellipses, took this idea—the infinitely distant point—one step further. Ellipses have two centers, or *foci;* the more elongated the ellipse, the farther apart these foci are. And all ellipses have the same property: if you had a mirror in the shape of an ellipse and you placed a lightbulb at one focus, all the beams of light would converge at the other focus, no matter how stretched-out the ellipses are (Figure 29).

In his mind Kepler stretched an ellipse out more and more, dragging one focus farther and farther away. Then Kepler imagined that the second focus was infinitely far away: the second focus was a point at infinity. All of a sudden the ellipse becomes a parabola, and all of the lines that con-

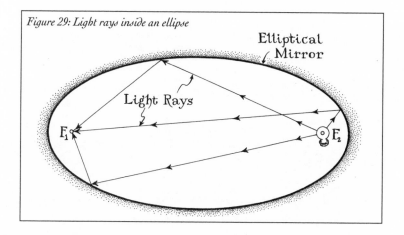

Figure 29: Light rays inside an ellipse

verged to a point become parallel lines. A parabola is simply
an ellipse with one focus at infinity (Figure 30).

You can see this very nicely with a flashlight. Go into a
dark room, stand next to a wall, and point the flashlight directly at it. You will get a nice, round circle of light projected
on the wall. Now slowly tilt the flashlight upward (Figure
31). You'll see the circle stretch out into an ellipse that gets
longer and longer as you increase the tilt. All of a sudden, the
ellipse opens up and becomes a parabola. Thus, Kepler's
point at infinity proved that parabolas and ellipses are actu-

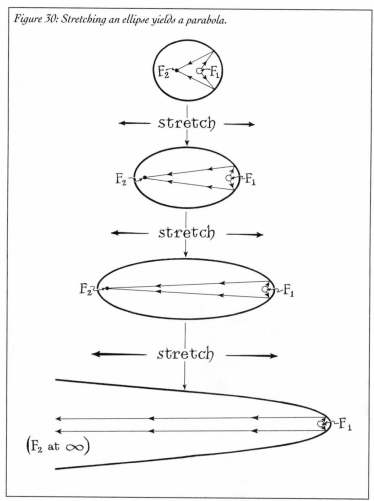

Figure 30: Stretching an ellipse yields a parabola.

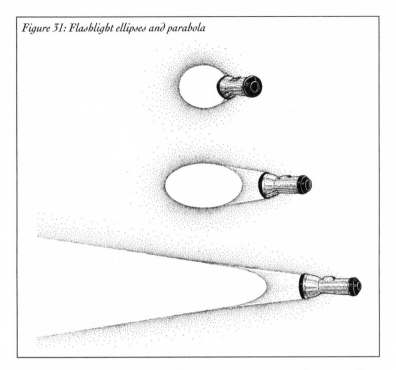

Figure 31: Flashlight ellipses and parabola

ally the same thing. This was the beginning of the discipline of projective geometry, where mathematicians look at the shadows and projections of geometric figures to uncover hidden truths even more powerful than the equivalence of parabolas and ellipses. However, it all depended upon accepting a point at infinity.

Gérard Desargues, a seventeenth-century French architect, was one of the early pioneers of projective geometry. He used the point at infinity to prove a number of important new theorems, but Desargues's colleagues couldn't understand his terminology and concluded that Desargues was nuts. Though a few mathematicians, like Blaise Pascal, picked up on Desargues's work, it was forgotten.

None of this mattered to Jean-Victor Poncelet. As Monge's student, Poncelet had learned the technique of projecting diagrams onto two planes, and as a prisoner of war he had a lot of spare time on his hands. He used his stay in prison to reinvent the concept of a point at infinity, and combining it with Monge's work, he became the first true projective geometer. Upon his return from Russia (carrying a Russian abacus, by then an archaic oddity) he raised the discipline to a high art.* However, Poncelet had no idea that projective geometry would reveal the mysterious nature of zero, because the second important advance, the complex plane, was still needed. We must turn to Germany for this piece of the puzzle.

Carl Friedrich Gauss, born in 1777, was a German prodigy, and he began his mathematical career with an investigation of imaginary numbers. His doctoral thesis was a proof of the fundamental theorem of algebra—proving that a polynomial of degree n (a quadratic has degree 2, a cubic has degree 3, a quartic has degree 4, and so on) has n roots. This is only true if you accept imaginary numbers as well as real numbers.

Throughout his life Gauss worked on an incredible variety of topics—his work on curvature would become a key component of Einstein's general theory of relativity—but it was Gauss's way of graphing complex numbers that revealed a whole new structure in mathematics.

In the 1830s Gauss realized that each complex number—numbers that have real and imaginary parts, like $1 - 2i$ —can be displayed on a Cartesian grid. The horizontal axis represents the real part of the complex number, while the vertical axis represents the imaginary part (Figure 32). This simple construction, called the complex plane, revealed a lot about the way numbers work. Take, for example, the number i. The angle between i and the x-axis is 90 degrees

---

* Poncelet's projective geometry brought about one of the oddest concepts in mathematics: the principle of *duality*. In high school geometry, you are taught that two points determine a line. But if you accept the idea of a point at infinity, two lines always determine a point. Points and lines are *dual* to each other. Every theorem in Euclidean geometry can be *dualized* in projective geometry, setting up a whole set of new theorems in the parallel universe of projective geometry.

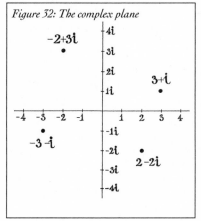

Figure 32: *The complex plane*

(Figure 33). What happens when you square i? Well, by definition, $i^2 = -1$—a point whose angle is 180 degrees from the x-axis; the angle has doubled. The number i^3 is equal to $-i$ — 270 degrees from the x-axis; the angle has tripled. The number $i^4 = 1$; we have gone around 360 degrees—exactly four times the original

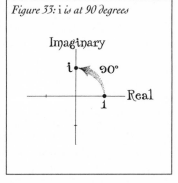

Figure 33: i *is at 90 degrees*

angle (Figure 34). This is not a coincidence. Take any complex number and measure its angle. Raising a number to the nth power multiplies its angle by n. And as you keep raising the number to higher and higher powers, the number will spiral inward or outward, depending on whether the number is on the inside or on the outside of the unit circle, a circle centered at the origin with radius 1 (Figure 35). Multiplication and exponentiation in the complex plane became geometric

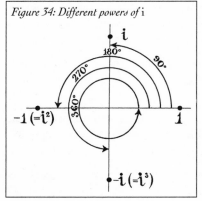

Figure 34: *Different powers of* i

ideas; you could actually see them happening. This was the second big advance.

The person who combined these two ideas was a student of Gauss's: Georg Friedrich Bernhard Riemann. Riemann merged projective geometry with the complex numbers, and all of a sudden lines became

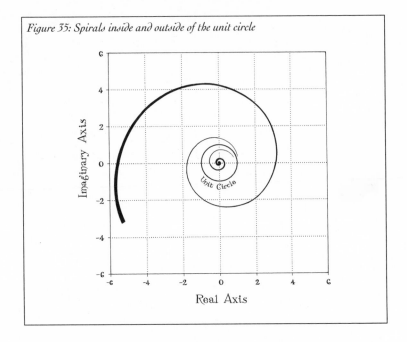

Figure 35: Spirals inside and outside of the unit circle

circles, circles became lines, and zero and infinity became the poles on a globe full of numbers.

Riemann imagined a translucent ball sitting atop the complex plane, with the south pole of the ball touching zero. If there were a tiny light at the north pole of the ball, any figures that are marked on the ball would cast shadows on the plane below. The shadow of the equator would be a circle around the origin. The shadow of the southern hemisphere is inside the circle and the shadow of the northern hemisphere is outside (Figure 36). The origin — zero — corresponds to the south pole. Every point on the ball has a shadow on the complex plane; in a sense, every point on the ball is equivalent to its shadow on the plane and vice versa. Every circle on the plane is the shadow of a circle on the ball, and a circle on the ball corresponds to a circle on the plane . . . with one exception.

If you've got a circle that goes through the north pole of the ball, the shadow is no longer a circle. It is a line. The north pole is like the point at infinity that Kepler and Poncelet

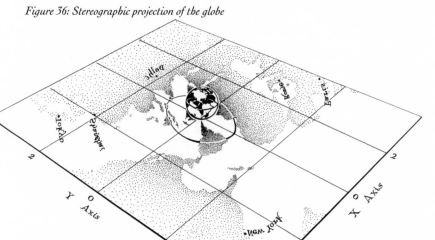

Figure 36: *Stereographic projection of the globe*

imagined. Lines on the plane are simply circles on the sphere that go through the north pole—the point at infinity (Figure 37).

Once Riemann saw that the complex plane (with a point at infinity) was the same thing as a sphere, mathematicians could see multiplication, division, and other, more difficult operations by analyzing the way the sphere deformed and rotated. For instance, multiplying by the number *i* was equivalent to spinning the sphere 90 degrees clockwise. If you

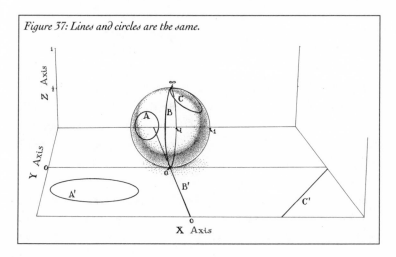

Figure 37: *Lines and circles are the same.*

take a number x and replace it with $(x - 1)/(x + 1)$, that is equivalent to rotating the whole globe by 90 degrees so that the north and south poles lie on the equator (Figures 38, 39, 40). Most interesting of all,

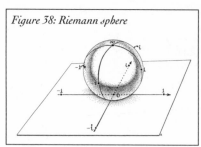

Figure 38: Riemann sphere

if you take a number x and replace it with its reciprocal $1/x$, that is equivalent to flipping the sphere upside down and reflecting it in a mirror. The north pole becomes the south pole and the south pole becomes the north pole: zero becomes infinity and infinity becomes zero. It's all built into the geometry of the sphere; $1/0 = \infty$ and $1/\infty = 0$. Infinity and zero are simply opposite poles on the Riemann sphere, and they can switch places in a blink. And they have equal and opposite powers.

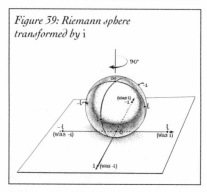

Figure 39: Riemann sphere transformed by i

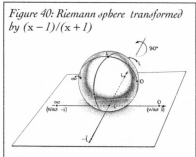

Figure 40: Riemann sphere transformed by $(x - 1)/(x + 1)$

Take all of the numbers in the complex plane and multiply them by two. That is like putting your hands on the south pole and stretching a rubber cover on the sphere away from the south pole and toward the north pole. Multiplying by one-half has the opposite effect. It is like stretching the rubber cover away from the north pole and toward the south pole. Multiplying by infinity is like sticking a needle in the south pole; the rubber sheet all flings upward toward the north pole: anything times infinity is infinity. Multiplying by zero is like sticking a needle on the north pole and everything winds up at zero: anything times zero is zero. Infinity and zero are equal and opposite — and equally destructive.

Zero and infinity are eternally locked in a struggle to engulf all the numbers. Like a Manichaean nightmare, the two sit on opposite poles of the number sphere, sucking numbers in like tiny black holes. Take any number on the plane. For the sake of argument, we'll choose $i/2$. Square it. Cube it. Raise it to the fourth power. The fifth. The sixth. The seventh. Keep multiplying. It slowly spirals toward zero like water down a drain. What happens to $2i$? The exact opposite. Square it. Cube it. Raise it to the fourth power. It spirals outward (Figure 41). But on the number sphere, the two curves are duplicates of each other; they are mirror images (Figure 42). All numbers in the complex plane suffer this fate. They are drawn inexorably toward 0 or toward ∞. The only numbers that escape are the ones that are equally distant from the two rivals — the numbers on the equator, like 1, -1, and i. These numbers, pulled by the tug of both zero and infinity, spiral around on the equator forever and ever, never able to escape the grasp of either. (You can see this on your calculator. Enter a number — any number. Square it. Square it again. Do it again and again; the number will quickly zoom toward infinity or toward zero, except if you entered 1 or -1 to begin with. There is no escape.)

The Infinite Zero

My theory stands as firm as a rock; every arrow directed against it will return quickly to its archer. How do I know this? I have studied it. . . . I have followed its roots, so to speak, to the first infallible cause of all created things.

— GEORG CANTOR

Infinity was no longer mystical; it became an ordinary number. It was a specimen impaled on a pin, ready for study, and

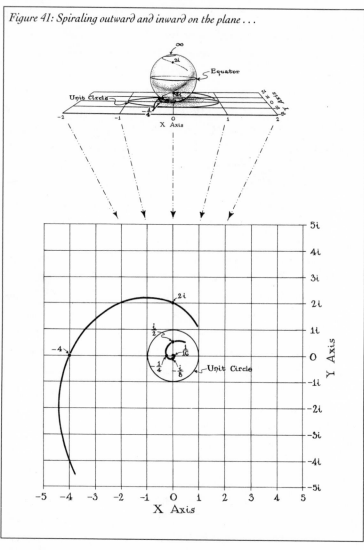

Figure 41: *Spiraling outward and inward on the plane . . .*

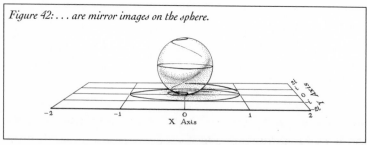

Figure 42: *. . . are mirror images on the sphere.*

mathematicians were quick to analyze it. But in the deepest infinity—nestled within the vast continuum of numbers—zero kept appearing. Most appalling of all, infinity itself can be a zero.

In the old days, before Riemann saw that the complex plane was really a sphere, functions like $1/x$ would stump mathematicians. When x goes to zero, $1/x$ gets bigger and bigger and bigger and finally just blows up and goes off to infinity. Riemann made it perfectly acceptable to go off to infinity; since infinity is just a point on the sphere like any other point, it was no longer something to be feared. In fact, mathematicians started analyzing and classifying the points where a function blows up: *singularities*.

The curve $1/x$ has a singularity at the point $x = 0$— a very simple sort of singularity that mathematicians dubbed a *pole*. There are other types of singularities as well; for instance, the curve $\sin(1/x)$ has an *essential* singularity at $x = 0$. Essential singularities are weird beasts; near a singularity of this sort, a curve goes absolutely berserk. It oscillates up and down faster and faster as it approaches the singularity, whipping from positive to negative and back again. In even the tiniest neighborhood around the singularity, the curve takes on almost every conceivable value over and over and over again. Yet as weird as these singularities behave, they were no longer mysterious to mathematicians, who were learning to dissect the infinite.

The master anatomist of the infinite was Georg Cantor. Though he was born in Russia in 1845, Cantor spent most of his life in Germany. And it was in Germany—the land of Gauss and of Riemann—where infinity's secrets were revealed. Unfortunately, Germany was also the land of Leopold Kronecker, the mathematician who would hound Cantor into a mental institution.

Underneath Cantor's conflict with Kronecker was a vision of the infinite, a vision that can be described with a simple puzzle. Imagine that there is a large stadium filled with people and you want to know whether there are more seats, more people, or an equal number of both. You could count the number of people and count the number of seats and then compare the two numbers, but that would take a lot of time. There's a much cleverer way. Just ask everyone to sit down in a seat. If there are empty seats, then there are too few people. If people remain standing, there are too few seats. If every seat is filled and nobody is left standing, then the number of people and seats are equal.

Cantor generalized this trick. He said that two sets of numbers are the same size when one set of numbers can "sit" on top of another set of numbers—one to a customer—with none left over. For instance, consider the set {1, 2, 3}. It is the same size as {2, 4, 6} because we can make a perfect seating pattern where all the numbers are "seated" and all the "seats" are occupied:

$$
\begin{array}{ccc}
1 & 2 & 3 \\
| & | & | \\
2 & 4 & 6
\end{array}
$$

But it's not the same size as {2, 4, 6, 8}

$$
\begin{array}{cccc}
1 & 2 & 3 & \\
| & | & | & | \\
2 & 4 & 6 & 8
\end{array}
$$

because 8 is an empty "seat."

Things get really interesting when you get to infinite sets. Consider the set of whole numbers: {0, 1, 2, 3, 4, 5, . . . }. Obviously, this set is equal to itself; we can just have each number "sit" upon itself:

$$
\begin{array}{ccccccc}
0 & 1 & 2 & 3 & 4 & 5 & \ldots \\
| & | & | & | & | & | & \\
0 & 1 & 2 & 3 & 4 & 5 & \ldots
\end{array}
$$

There's no trick here. Every set is obviously equal to itself. But what happens when we start removing numbers from the set? For instance, what happens when we remove 0? Oddly enough, removing 0 doesn't change the size of the set at all. By rearranging the seating pattern slightly, we can ensure that everybody has a seat, and all the seats are still taken:

1	2	3	4	5	6	...
|	|	|	|	|	|	
0	1	2	3	4	5	...

The set is the same size, even though we've removed something from it. In fact, we can remove an infinite number of elements from the set of whole numbers—we can delete the odd numbers, for example—and the size of the set remains unchanged. Everybody still has a seat, and every seat is filled:

0	2	4	6	8	10	...
|	|	|	|	|	|	
0	1	2	3	4	5	...

This is the definition of the infinite: it is something that can stay the same size even when you subtract from it.

The even numbers, the odd numbers, the whole numbers, the integers—all of these sets were the same size, a size that Cantor soon dubbed $\aleph_0$ (aleph nought, named after the first letter of the Hebrew alphabet). Since these numbers are the same size as the counting numbers, any set of size $\aleph_0$ is called *countable*. (Of course, you couldn't really count them unless you had an infinite amount of time on your hands.) Even the rational numbers—the set of numbers that can be written as a/b for integers a and b—were countable. By a clever way of assigning rational numbers to their proper seats, Cantor showed that the rationals were an $\aleph_0$-sized set (see appendix D).

But as Pythagoras knew, the rationals aren't everything under the sun; both the rationals and irrationals make up the so-called real numbers. Cantor discovered that the set of real numbers is much, much bigger than the rationals. His proof was a very simple one.

Imagine that we've already got a perfect seating plan for the real numbers: every real number has a seat, and every seat is filled. That means we can make a list of seats, showing a seat's number along with the real number that is sitting in it. For instance, our list might look like the following:

Seat	Real Number
1	.3125123 . . .
2	.7843122 . . .
3	.9999999 . . .
4	.6261000 . . .
5	.3671123 . . .
etc.	etc.

The trick came when Cantor created a real number that was not on the list.

Look at the first digit of the first number on the list; in our example, it's a 3. If our new number were equal to the first number on the list, it would also have a first digit of 3 — but we can easily prevent that from happening. Let's just say that our new number has a first digit of 2. Since the first number on the list starts with a 3 and our new number starts with a 2, we know that the two numbers are different. (This is not strictly true. The number 0.300000 . . . is equal to 0.2999999 . . . , since there are two ways to write many rational numbers. But this is a minor point that is easily overcome. For the sake of clarity, we'll ignore that exception.)

On to the second number. How can we make sure that our new number is different from the second number on the list? Well, we've already determined the first digit in our new number, so we can't pull exactly the same trick, but we can do something just as good. The second number on the list has an 8 for its second digit. If our new number has a 7 for its second digit, we can ensure that our new number is not the same as the second number on the list; their second digits don't match, so they aren't the same thing. We do the same thing on down the list; look at the third digit of the third number and change it, look at the fourth digit on the fourth number and change it, and so on.

Seat	Real Number	
1	.③125123 . . .	Our new number's first digit is 2 (different from 3).
2	.7⑧43122 . . .	Our new number's second digit is 7 (different from 8).
3	.99⑨9999 . . .	Our new number's third digit is 8 (different from 9).
4	.626①000 . . .	Our new number's fourth digit is 0 (different from 1).
5	.3671①23 . . .	Our new number's fifth digit is 0 (different from 1).
etc.	etc.	etc.

Yielding a new number, .27800 . . . , that

is different from the first number (their first digits don't match),

is different from the second number (their second digits don't match),

is different from the third number (their third digits don't match),

is different from the fourth number (their fourth digits don't match),

and so forth.

Going down the diagonal in this way, we create a new number. This process ensures that it's different from all the other numbers on the list. If it is different from all the numbers on the list, it can't be on the list—but we already assumed our list contains all real numbers; after all, it was a perfect seating list. This is a contradiction. The perfect seating list cannot exist.

The real numbers are a bigger infinity than the rational numbers. The term for this type of infinity was $\aleph_1$, the first *uncountable* infinity. (Technically, the term for the infinity of the real line was C, or the continuum infinity. For years mathematicians struggled to determine whether C was indeed $\aleph_1$. In 1963 a mathematician, Paul Cohen, proved that this puzzle, the so-called continuum hypothesis, was neither provable nor disprovable, thanks to Gödel's incompleteness theorem. Today most mathematicians accept the continuum hypothesis as true, though some study *non-Cantorian transfinite numbers* where the continuum hypothesis is taken to be false.) In Cantor's mind there were an infinite number of infinities—the transfinite numbers—each nested in the other. $\aleph_0$ is smaller than $\aleph_1$, which is smaller than $\aleph_2$, which is smaller than $\aleph_3$, and so forth. At the top of the chain sits the ultimate infinity that engulfs all other infinities: God, the infinity that defies all comprehension.

Unfortunately for Cantor, not everyone had the same vision of God. Leopold Kronecker was an eminent professor at the University of Berlin, and one of Cantor's teachers. Kronecker believed that God would never allow such ugliness as the irrationals, much less an ever-increasing set of Russian-doll infinities. The integers represented the purity of God, while the irrationals and other bizarre sets of numbers were abominations—figments of the imperfect human mind. Cantor's transfinite numbers were the worst of the lot.

Disgusted with Cantor, Kronecker launched vitriolic attacks against Cantor's work and made it extremely difficult for him to publish papers. When Cantor applied for a position at the University of Berlin in 1883, he was rejected; he had to settle for a professorship at the much less prestigious University of Halle instead. Kronecker, who was influential at Berlin, was likely to blame. The same year, he wrote a defense against Kronecker's attacks. Then, in 1884, the depressed Cantor had his first mental breakdown.

It would be little comfort to Cantor that his work was the foundation of a whole new branch of mathematics: set theory. Using set theory, mathematicians would not only create the numbers we know out of nothing at all, they would create numbers that were previously unheard of—infinite infinities that can be added to, multiplied with, subtracted from, and divided by other infinities, just like ordinary numbers. Cantor opened up a whole new universe of numbers. The German mathematician David Hilbert would say, "No one shall expel us from the paradise which Cantor has created for us." But it was too late for Cantor. Cantor was in and out of mental institutions for the remainder of his life, and he died in the mental hospital at Halle in 1918.

In the battle between Kronecker and Cantor, Cantor would ultimately prevail. Cantor's theory would show that Kronecker's precious integers—and even the rational numbers—were nothing at all. They were an infinite zero.

There are an infinite number of rationals, and between any two numbers you choose, no matter how close together, there are still an infinite number of rationals. They are everywhere. But Cantor's hierarchy of infinities would tell a different tale: it would show just how little space the rational numbers take up on the number line.

It takes a clever trick to do such an intricate calculation. Irregularly shaped objects can be very difficult to measure. For instance, imagine that you've got a stain on your wood floor. How much area does the stain take up? It's not so obvious. If the stain were shaped like a circle, or like a square or a triangle, it would be easy to figure out; just take a ruler and measure its radius or its height and base. But there's no formula for figuring out the area of an amoeba-shaped mess. However, there is another way.

Take a rectangular carpet and place it on top of the stain. If the carpet covers the stain entirely, we know that the stain is smaller than the carpet; if the carpet is one square foot, then the stain must take up less than one square foot. If we use smaller carpets, our approximation gets better and better. Perhaps the stain is covered by five carpets of size one-eighth square foot; we would then know that the stain takes up at most five-eighths of a square foot, which is less than our approximation with a one-square-foot carpet. As you make the carpets smaller and smaller, the covering gets better and better, and your total carpet area approaches the true size of the stain; in fact, you can define the size of the stain as the limit as your carpets approach zero size (Figure 43).

Let's do the same thing with the rational numbers—but this time our carpets are sets of numbers. For instance, the number 2.5 is "covered" by a carpet that includes, for example, all the numbers between 2 and 3—a carpet of size 1. Using this sort of carpet to cover the rational numbers has some very odd consequences, as Cantor soon showed, thanks to his seating chart. That seating chart accounts for all the rational numbers—it assigns each of them a seat—so we can count them off one by one, in order, based on their seat number. Take the first rational number and imagine it on the number line. Let's cover it with a carpet of size 1. Lots of other numbers are covered by that carpet, but we don't have to worry about that. So long as the first number is covered, we are happy.

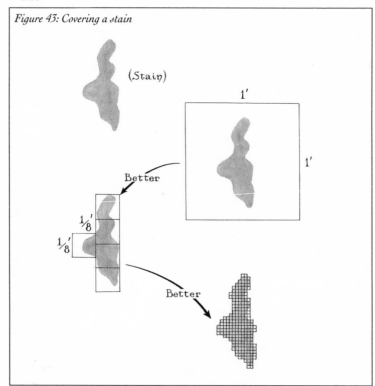

Figure 43: Covering a stain

Now take the second number. Cover it with a carpet of size 1/2. Take the third number and cover it with a carpet of size 1/4, and so forth. Go on and on to infinity; since every rational number is on the seating chart, every rational number will eventually be covered by a carpet. What is the total size of the carpets? It's our old friend, the Achilles sum. Adding up the size of the carpets, we see $1 + 1/2 + 1/4 + 1/8 + \ldots + 1/2^n$ goes to 2 as n goes to infinity. So we can cover the infinite cohorts of rational numbers in the number line with a set of carpets, and the total size of the carpets is 2. This means that the rational numbers take up less than two units of space.

Just as we did with the stain, let's make the carpet sizes even smaller to get a better approximation of the size of the rationals. Instead of starting with a carpet of size 1, starting with a carpet of size 1/2 makes the total size of the carpets equal to 1; the rational numbers take up less than one unit of space, in total. If we start off with an initial carpet that has size 1/1000, all the carpets, in total, take up less than 1/500 unit of space; all the rational numbers take up less room than 1/500 unit. If we start with a carpet the size of half an atom, we can cover all the rational numbers on the number line with carpets that, in total, take up less room than an atom. Yet even those tiny carpets, all of which can fit in the span of an atom, cover all of the rational numbers (Figure 44).

We can get smaller and smaller — as small as we want. We can cover the rationals with carpets that, summed together, fit in the size of half an atom — or a neutron — or a quark — or as small as we can possibly imagine.

How big are the rational numbers, then? We defined size as a limit — the sum of the carpets as the individual sizes go to zero. Yet at the same time, we saw that as the carpets get smaller and smaller, the sum of the cover gets tinier and

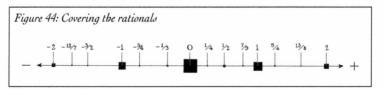

Figure 44: Covering the rationals

tinier—smaller than an atom or a quark or a millionth-billionth part of a quark—and we can still cover the rationals. What is the limit of something that gets smaller and smaller and smaller without stopping?

Zero.

How big are the rational numbers? They take up no space at all. It's a tough concept to swallow, but it's true.

Even though there are rational numbers everywhere on the number line, they take up no space at all. If we were to throw a dart at the number line, it would *never* hit a rational number. Never. And though the rationals are tiny, the irrationals aren't, since we can't make a seating chart and cover them one by one; there will always be uncovered irrationals left over. Kronecker hated the irrationals, but they take up *all* the space in the number line.

The infinity of the rationals is nothing more than a zero.

Chapter **7**

Absolute Zeros

[THE PHYSICS OF ZERO]

Sensible mathematics involves neglecting a quantity
when it is small — not neglecting it because it is infinitely
great and you do not want it!

— P. A. M. Dirac

It was finally unmistakable: infinity and zero are inseparable and are essential to mathematics. Mathematicians had
no choice but to learn to live with them. For physicists, however, zero and infinity seemed utterly irrelevant to the workings of the universe. Adding infinities and dividing by zeros
might be a part of mathematics, but it is not the way of nature.

Or so scientists had hoped. As mathematicians were uncovering the connection between zero and infinity, physicists
began to encounter zeros in the natural world; zero crossed
over from mathematics to physics. In thermodynamics a zero
became an uncrossable barrier: the coldest temperature possible. In Einstein's theory of general relativity, a zero became

a black hole, a monstrous star that swallows entire suns. In quantum mechanics, a zero is responsible for a bizarre source of energy—infinite and ubiquitous, present even in the deepest vacuum—and a phantom force exerted by nothing at all.

Zero Heat

When you can measure what you are speaking about, and express it in numbers, you know something about it; but when you cannot measure it, when you cannot express it in numbers, your knowledge is of a meager and unsatisfactory kind: it may be the beginning of knowledge, but you have scarcely, in your thoughts, advanced to the stage of science.

— WILLIAM THOMSON, LORD KELVIN

The first inescapable zero in physics comes from a law that had been in use for half a century. This law was discovered in 1787 by Jacques-Alexandre Charles, a French physicist already famous for being the first to fly aboard a hydrogen balloon. Charles isn't remembered for his aeronautic stunts, but for the law of nature that bears his name.

Charles, like many physicists of his time, was fascinated with the very different properties of gases. Oxygen makes embers burst into flame, while carbon dioxide snuffs them out. Chlorine is green and is deadly; nitrous oxide is colorless and makes people giggle. Yet all these gases have very basic properties in common: heat them up and they expand; cool them down and they contract.

Charles discovered that this behavior is extremely regular and predictable. Take an equal volume of any two different gases and put them in identical balloons. Heat them up by the same amount and they expand by the same amount; cool them down together and they contract in unison. Furthermore, for each degree up or down you go, you gain or lose a certain percentage of the volume. Charles' law describes the relationship of the volume of a gas to its temperature.

In the 1850s, however, William Thomson, a British physicist, noticed something odd about Charles' law: the specter of zero. Lower the temperature and the volume of the balloons gets smaller and smaller. Keep lowering at a steady pace and the balloons keep shrinking at a constant rate, but they cannot go on shrinking forever. There is a point at which gas, in theory, takes up no space at all; Charles' law says that a balloon of gas must shrink to zero space. Of course, zero space is the smallest possible volume; when a gas reaches this point, it takes up no space at all. (It certainly can't take up negative space.) If the volume of a gas is related to its temperature, a minimum volume means that there is a minimum temperature. A gas cannot keep getting colder and colder indefinitely; when you can't shrink the balloon any further, you can't lower the temperature any further. This is *absolute zero*. It is the lowest temperature possible, a little more than 273 degrees Celsius below the freezing point of water.·

Thomson is better known as Lord Kelvin, and it is for Kelvin that the universal temperature scale is named. In the centigrade scale, zero degrees is the freezing point of water. In the Kelvin scale, zero degrees is absolute zero.

Absolute zero is the state where a container of gas has been drained of all of its energy. This is, in actuality, an unattainable goal. You can never cool an object to absolute zero. You can get very close; thanks to laser cooling, physicists can chill atoms to a few millionths of a degree above the ultimate coldness. However, everything in the universe is conspiring to stop you from actually reaching absolute zero. This is because any object that has energy is bouncing around—and radiating light. For instance, people are made up of molecules of water and a few organic contaminants. All of these atoms are wiggling about in space; the higher the temperature, the faster the atoms wiggle. These wiggling atoms bump into one another, getting their neighbors to wiggle as well.

Say you are trying to cool a banana to absolute zero. To get rid of all of the energy in the banana, you've got to stop its atoms from moving around; you have to put it in a box and cool it down. However, the box the banana is in is made of atoms, too. The box's atoms are wiggling around, and they will bump the banana's atoms and set them in motion again. Even if you get the banana to float in a perfect vacuum in the center of the box, you can't stop the wiggling entirely, because dancing particles give off light. Light is constantly coming off of the box and striking the banana, getting the banana's molecules to move again.

All of the atoms that make up a tweezer, a refrigerator coil, and a tub of liquid nitrogen are moving and radiating, so the banana is constantly absorbing energy from the wiggles and radiation of the box it is in, from the tweezers you use to manipulate the banana, and from the refrigerator coil you use to cool it down. You cannot shield the banana from the box or the tweezer or the coil; the shield, too, is wiggling and radiating. Every object is influenced by the environment it's in, so it's impossible to cool anything in the universe—a banana, an ice cube, a dollop of liquid helium—to absolute zero. It is an unbreakable barrier.

Absolute zero was a discovery that had a very different flavor from Newton's laws. Newton's equations gave physicists power. They could predict the orbits of the planets and the motion of objects with great accuracy. On the other hand, Kelvin's discovery of absolute zero told physicists what they *couldn't* do. They couldn't ever reach absolute zero. This barrier was disappointing news to the physics world, but it was the beginning of a new branch of physics: thermodynamics.

Thermodynamics is the study of the way heat and energy behave. Like Kelvin's discovery of absolute zero, the laws of thermodynamics erected impenetrable barriers that no scientists can ever cross, no matter how hard they try. For instance, thermodynamics tells you that it is impossible to create a perpetual-motion machine. Avid inventors tend to swamp physics departments and science magazines with blueprints for incredible machines—machines that eternally generate power without any source of energy. However, the laws of thermodynamics state that it is impossible to create such a machine. It is another task that cannot be done, no matter how hard you try. It is impossible even to get a machine to run without wasting energy, frittering some of its power into the universe as heat. (Thermodynamics is worse than a casino; you can't win, no matter how much you work at it. You can't even break even.)

From thermodynamics came the discipline of statistical mechanics. By looking at the collective motion of groups of atoms, physicists could predict the way matter behaves. For instance, the statistical description of a gas explains Charles' law. As you raise the temperature of a gas, the average molecule moves faster and smashes harder into the walls of its container. The gas pushes harder on the walls: the pressure goes up. Statistical mechanics—the theory of wiggles—explained some of the basic properties of matter, and it even seemed to explain the nature of light itself.

The nature of light was a problem that had consumed scientists for centuries. Isaac Newton believed that light was composed of little particles that flowed from every bright object. Over time, though, scientists came to believe that light was not in fact a particle, but a wave. In 1801 a British scientist discovered that light interferes with itself, apparently putting the matter to rest once and for all.

Interference happens with all sorts of waves. When you drop a stone into a pond, you create circular ripples in the water—waves. The water bobs up and down, and crests and troughs spread outward in a circular pattern. If you drop two stones at the same time, the ripples interfere with one another. You can see this more clearly if you dip two oscillating pistons into a tub of water. When a crest from one piston runs into a trough from the other, the two cancel out; if you look carefully at the pattern of ripples, you can see lines of still, wave-free water (Figure 45).

The same thing is true of light. If light shines through two small slits, there are areas that are dark—wave-free (Figure 46). (You can see a related effect at home. Hold your fingers together; you should have tiny gaps where some light can get through. Gaze through one of those gaps at a lightbulb and you'll see faint dark lines, especially near the top and bottom of the gap. These lines, too, are due to the wavelike nature of light.) Waves interfere in this way; particles do not. Thus, the phenomenon of interference seemed to settle the question of light's nature once and for all. Physicists concluded that light was not a particle, but a wave of electric and magnetic fields.

This was the state of the art in the mid-1800s, and it seemed to mesh perfectly with the laws of statistical mechan-

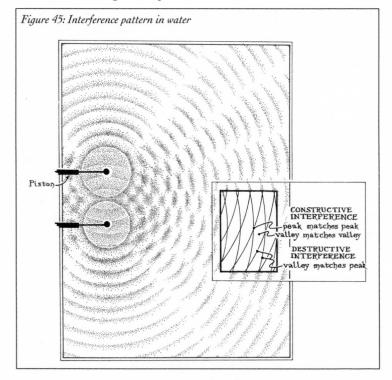

Figure 45: Interference pattern in water

Piston

CONSTRUCTIVE INTERFERENCE
peak matches peak
valley matches valley

DESTRUCTIVE INTERFERENCE
valley matches peak

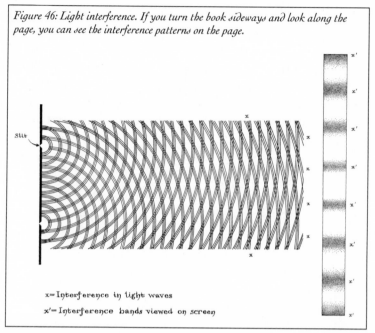

Figure 46: Light interference. If you turn the book sideways and look along the page, you can see the interference patterns on the page.

x = Interference in light waves

x' = Interference bands viewed on screen

ics. Statistical mechanics tells you how the molecules of matter wiggle; the wave theory of light implied that these molecular wiggles somehow cause ripples of radiation—light waves. Better yet, the hotter the object is, the faster its molecules move; at the same time, the hotter the object, the more energetic the ripples of light it sends out. This works out perfectly. With light, the faster the wave bobs up and down—the higher its frequency—the more energy it has. (Also, the higher its frequency, the smaller its *wavelength:* the distance between two wave crests.) Indeed, one of the most important thermodynamic laws—the so-called Stefan-Boltzmann equation—seems to tie the wiggles of molecules to the wiggles of light. It relates the temperature of an object to the total amount of light energy it radiates. This was the biggest victory for statistical mechanics and the wave theory of light. (The equation states that the radiated energy is proportional to the temperature raised to the fourth power. It not only tells how much radiation an object gives off, but also how hot an object gets when irradiated with a given amount of energy. This is the law that physicists used—along with a passage in the book of Isaiah—to determine that heaven is more than 500 degrees Kelvin.)

Unfortunately, the victory would not last for long. At the turn of the century, two British physicists tried to use the wiggle theory to solve a simple problem. It was a fairly straightforward calculation: how much light does an empty cavity radiate? Applying the basic equations of statistical mechanics (which tells how the molecules wiggle) and the equations that describe the way electric and magnetic fields interact (which tells how light wiggles), they came up with an equation that describes what wavelengths of light a cavity radiates at any given temperature.

The so-called Rayleigh-Jeans law, named after the physicists Lord Rayleigh and Sir James Jeans, worked fairly well. It did a good job of predicting the amount of large-wavelength, low-energy light that comes off a hot object. At high energies, though, the equation faltered. The Rayleigh-Jeans law predicted that an object gives off more and more light at smaller and smaller wavelengths (and thus higher and higher energies). Consequently at realms close to zero wavelength, the object gives off an *infinite* amount of high-energy light. According to the Rayleigh-Jeans equation, every object is constantly radiating an infinite amount of energy, no matter what its temperature is; even an ice cube would be radiating enough ultraviolet rays, x rays, and gamma rays to vaporize everything around. This was the "ultraviolet catastrophe." Zero wavelength equals infinite energy; zero and infinity conspired to break a nice, neat system of laws. Solving this paradox quickly became the leading puzzle in physics.

Rayleigh and Jeans had done nothing wrong. They used equations that physicists thought were valid, manipulated them in an accepted way, and came out with an answer that didn't reflect the way the world works. Ice cubes don't wipe out civilizations with bursts of gamma rays, though following the then-accepted rules of physics led inexorably to that conclusion. One of the laws of physics had to be wrong. But which one?

The Quantum Zero: Infinite Energy

To physicists, vacuum has all particles and forces latent in it. It's a far richer substance than the philosopher's nothing.

— SIR MARTIN REES

The ultraviolet catastrophe led to the quantum revolution. Quantum mechanics got rid of the zero in the classical theory of light—removing the infinite energy that supposedly came from every bit of matter in the universe. However, this was not much of a victory. A zero in quantum mechanics means that the entire universe—including the vacuum—is filled with an infinite amount of energy: the *zero-point* energy. This, in turn, leads to the most bizarre zero in the universe: the phantom force of nothing.

In 1900, German experimenters tried to shed some light on the ultraviolet catastrophe. By making careful measurements of how much radiation came off objects at various temperatures, they showed that the Rayleigh-Jeans formula was, indeed, failing to predict the true amount of light that comes from objects. A young physicist named Max Planck looked at the new data and within hours came up with a new equation that replaced the Rayleigh-Jeans formula. Not only did Planck's formula explain the new measurements, it solved the ultraviolet catastrophe. The Planck formula did not zoom off to infinity as the wavelength decreased; instead of having the energy get bigger and bigger as the wavelength goes down, it got smaller and smaller again (Figure 47). Unfortunately,

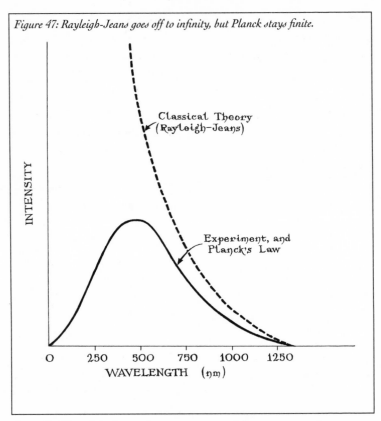

Figure 47: Rayleigh-Jeans goes off to infinity, but Planck stays finite.

Classical Theory
(Rayleigh-Jeans)

Experiment, and
Planck's Law

INTENSITY

O   250   500   750   1000   1250
WAVELENGTH   (ŋm)

though Planck's formula was correct, its repercussions were more troubling than the ultraviolet catastrophe it solved.

The problem arose because the ordinary assumptions of statistical mechanics—the laws of physics—did not lead Planck to his formula. The laws of physics had to change to accommodate the Planck formula. Planck later described what he did as an "act of desperation"; nothing less than desperation would compel a physicist to make such a seemingly nonsensical change in the laws of physics: According to Planck, molecules are forbidden to move in most ways. They vibrate only with certain acceptable energies, called quanta. It is impossible for molecules to have energies in between these acceptable values.

This might not seem like such a strange assumption, but

it is not the way the world seems to work. Nature doesn't move in jumps. It would seem silly to have five-foot-tall people and six-foot-tall people but nothing in between. It would be ridiculous if cars drove at 30 miles an hour and 40 miles an hour, but never at 33 or 38 miles an hour. However, a quantum car would behave in exactly this way. You might be driving along at 30 miles an hour, but when you step on the gas, all of a sudden you would instantly—pop!—be driving 40 miles an hour. Nothing in between is allowed, so to get from 30 to 40 miles an hour, you have to make a *quantum leap.* In the same way, quantum people could not grow very easily; they would hover at four feet for a number of years, and then, in a fraction of a second—pop!—they would be five feet tall. The quantum hypothesis violates everything our everyday experience tells us.

Even though it doesn't agree with the way nature seems to work, Planck's strange hypothesis—that molecular vibrations were *quantized*—led to the correct formula for the frequencies of light that come off an object. Even though physicists quickly realized that Planck's equation was right, they did not accept the quantum hypothesis. It was too bizarre to accept.

An unlikely candidate would turn the quantum hypothesis from a peculiarity to an accepted fact. Albert Einstein, a twenty-six-year-old patent clerk, showed the physics world that nature worked in quanta rather than in smooth increments. He would later become the chief opponent of the theory he helped create.

Einstein didn't seem like a revolutionary. When Max Planck was turning the physics world on its head, Albert Einstein was scrambling for a job. Out of money, he took a temporary position at the Swiss patent office, a far cry from the assistantship at a university that he wanted. By 1904 he was married, had a newborn son, and was laboring in the patent office—hardly the path to greatness. However, in March 1905, he wrote a paper that would eventually earn him the

Nobel Prize. This paper—which explained the *photoelectric effect*, brought quantum mechanics into the mainstream. Once quantum mechanics was accepted, so, too, would the mysterious powers of zero.

The photoelectric effect was discovered in 1887 when the German physicist Heinrich Hertz discovered that a beam of ultraviolet light could cause a plate to spark: electrons quite simply pop out of the metal when light shines on it. This phenomenon, causing sparks with a beam of light, was very puzzling to classical physicists. Ultraviolet light is light with a lot of energy, so scientists naturally concluded that it took quite a bit of energy to kick an electron out of an atom. But according to the wave theory of light, there is another way to get high-energy light: make it brighter. A very bright blue light, for instance, might have as much energy as a dim ultraviolet beam; therefore, a bright blue light should be able to kick electrons out of atoms, just as a dim ultraviolet beam can.

This simply is not the case, as experiments quickly showed. Even a dim beam of ultraviolet (high frequency) light causes electrons to get knocked out of the metal. However, if you lower the frequency just a little bit beyond a critical threshold—making the light a wee bit too red—the sparking stops all of a sudden. No matter how bright the beam is, if the light is the wrong color, all the electrons in the metal stay put; none of them can escape. That's not the sort of thing a light wave would do.

Einstein solved this quandary—the puzzle of the photoelectric effect—but his solution was even more revolutionary than Planck's hypothesis. While Planck proposed that molecules' vibrations were quantized, Einstein proposed that light came in little packets of energy called photons. This idea conflicted with the accepted physics of light, because it meant that light was not a wave.

On the other hand, if light energy is bundled into little packets, then the photoelectric effect is easy to explain. The light is acting like little bullets that get shot into the metal.

When a bullet hits an electron, it gives it a nudge. If the bullet has enough energy—if its frequency is high enough—then it knocks the electron free. On the other hand, if a light particle doesn't have enough energy to nudge the electron out, then the electron stays put; the photon skitters away instead.

Einstein's idea explained the photoelectric effect brilliantly. Light is quantized into photons, directly contradicting the wave theory of light that had not been questioned for more than a century. Indeed, it turns out that light has *both* a wave nature and a particle nature. Though light acts like a particle sometimes, it acts like a wave at other times. In truth, light is neither particle nor wave, but a strange combination of the two. It's a hard concept to grasp. However, this idea is at the heart of the quantum theory.

According to quantum theory, everything—light, electrons, protons, small dogs—have both wavelike and particlelike properties. But if objects are particles and waves at the same time, what on earth could they be? Mathematicians know how to describe them: they are *wave functions*, solutions to a differential equation called the Schrödinger equation. Unfortunately, this mathematical description has no intuitive meaning; it is all but impossible to visualize what these wave functions are.* Worse yet, as physicists discovered the intricacies of quantum mechanics, stranger and stranger things began to appear. Perhaps the weirdest of all is caused by a zero in the equations of quantum mechanics: the zero-point energy.

---

* One thing that sometimes helps is thinking of the wave function (technically, the square of the wave function) as a measure of the probability about where a particle will be. An electron, say, is smeared out across space, but when you make a measurement to determine where it is, the wave function determines how likely you are to spot the electron at any given point in space. This very smeariness of nature was what Einstein objected to. His famous statement, "God does not play dice with the universe," was a rejection of the probabilistic way that quantum mechanics works. Unfortunately for Einstein, the laws of quantum mechanics work incredibly well, and you can't successfully explain quantum effects with traditional classical physics.

This strange force is woven into the mathematical equations of the quantum universe. In the mid-1920s a German physicist, Werner Heisenberg, saw that these equations had a shocking consequence: uncertainty. The force of nothing is caused by the Heisenberg uncertainty principle.

The concept of uncertainty pertains to scientists' ability to describe the properties of a particle. For instance, if we want to find a particular particle, we need to determine the particle's position and velocity—where it is and how fast it is going. Heisenberg's uncertainty principle tells us that we can't do even this simple act. No matter how hard we try, we cannot measure a particle's position and its velocity with perfect accuracy at the same time. This is because the very act of measuring destroys some of the information we are trying to gather.

To measure something, you need to prod it. For instance, imagine that you are measuring the length of a pencil. You could run your fingers along it and measure how long it is; however, you'll probably give the pencil a nudge, disturbing the pencil's velocity slightly. A better way would be to place a ruler gently next to the pencil, but in fact, comparing the lengths of the two objects also changes the pencil's speed a tiny bit. You can only look at the pencil when light is bouncing off it; though the disturbance is very slight, the photons that carom off the pencil nudge it ever so gently, changing the pencil's velocity a tiny bit. No matter what way you think of to measure the pencil, you will give it a tiny nudge in the process. Heisenberg's uncertainty principle shows that there is no possible way to measure the pencil's length—or an electron's position—and its velocity with perfect accuracy at the same time. In fact, the better you know a particle's position, the less you know about its velocity, and vice versa. If you measure an electron's position with zero error—you know exactly where it is at a given moment—you must have zero information about how fast it is going. And if you know a particle's velocity with infinite precision—zero error—you have infinite error when you measure its position; you know nothing at all about where it is.* You can never know both at the same time, and if you have some information about one, you must have some uncertainty about the other. It's another unbreakable law.

Heisenberg's uncertainty principle applies to more than just measurements performed by humans. Like the laws of thermodynamics, the principle applies to nature itself. Uncertainty makes the universe seethe with infinite energy. Imagine an extremely tiny volume in space, like a really small box. If we analyze what is going on inside that box, we can make some assumptions. For instance, we know, with some precision, the position of the particles inside. After all, they can't be outside the box; we know that they are restricted to a certain volume, because if they were outside the box, we would not be looking at them. Because we have some information about the particles' position, the Heisenberg uncertainty principle implies that we must have some uncertainty about the particles' velocity—their energy. As we make that box smaller and smaller, we know less and less about the particles' energy.

This argument holds everywhere in the universe—in the center of the earth and in the deepest vacuum of space. This means that in a sufficiently small volume, even in a vacuum, we have some uncertainty about the amount of energy inside. But uncertainty about the energy in a vacuum sounds ridiculous. The vacuum, by definition, has nothing in it—no particles, no light, nothing. Thus, the vacuum should have no energy at all. Yet according to Heisenberg's principle, we cannot know how much energy there is in a volume of the vacuum at any given time. The energy in a tiny volume of vacuum must be fluctuating constantly.

---

* To be precise, the Heisenberg uncertainty principle deals not with a particle's velocity but with momentum, which combines speed, direction, and information about the particle's mass. However, in this context, momentum, velocity, and even energy can be used almost interchangeably.

How could the vacuum, which has nothing in it, have any energy at all? The answer comes from another equation: Einstein's famous $E = mc^2$. This simple formula relates mass and energy: the mass of an object is equivalent to a certain amount of energy. (In fact, particle physicists don't measure the mass of the electron, say, in kilograms or pounds or any of the usual units of mass or weight. They say that the electron's *rest mass* is 0.511 MeV [million electron volts]—a lump of energy.) The fluctuation in the energy in the vacuum is the same thing as a fluctuation in the amount of mass. Particles are constantly winking in and out of existence, like tiny Cheshire cats. The vacuum is never truly empty. Instead, it is seething with these *virtual particles;* at every point in space, an infinite number are happily popping up and disappearing. This is the zero-point energy, an infinity in the formulas of quantum theory. Interpreted strictly, the zero-point energy is limitless. According to the equations of quantum mechanics, more power than is stored in all the coal mines, oil fields, and nuclear weapons in the world is sitting in the space inside your toaster.

When an equation has an infinity in it, physicists usually assume that there is something wrong; the infinity has no physical meaning. The zero-point energy is no different; most scientists ignore it completely. They simply pretend that the zero-point energy is zero, even though they know it is infinite. It's a convenient fiction, and it usually doesn't matter. However, sometimes it does. In 1948 two Dutch physicists, Hendrick B. G. Casimir and Dik Polder, first realized that the zero-point energy can't always be ignored. The two scientists were studying the forces between atoms when they realized that their measurements didn't match the forces that had been predicted. In a search for an explanation, Casimir realized that he had felt the force of nothing.

The secret to the Casimir force lies with the nature of waves. In ancient Greece, Pythagoras saw the peculiar behavior of waves that traveled up and down a plucked string— how certain notes were allowed and others were forbidden.

When Pythagoras strummed a string, the string sounded a clear note, the tone known as the fundamental. When he gently placed his finger in the middle of the string and plucked again, he got another nice, clear note, this time one octave above the fundamental. One-third of the way down yielded another nice tone. But Pythagoras realized that not all notes are allowed. When he placed his finger randomly on the string, he seldom got a clear note. Only certain notes can be played on the string; most are excluded (Figure 48).

Matter waves are not so different from string waves. Just as a guitar string of a given size is not capable of playing every possible note—some waves are "forbidden" from appearing on the string—some particle waves are forbidden from being inside a box. Put two metal plates close together, for instance, and you can't fit every sort of particle inside. Only those whose waves match the size of the box are allowed in (Figure 49).

Casimir realized

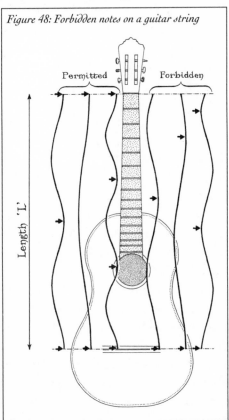

Figure 48: Forbidden notes on a guitar string

that the forbidden particle waves would affect the zero-point energy of the vacuum, since particles are everywhere winking in and out of existence. If you put two metal plates close together and some of those particles aren't allowed between the plates, then there are more particles on the outside of the

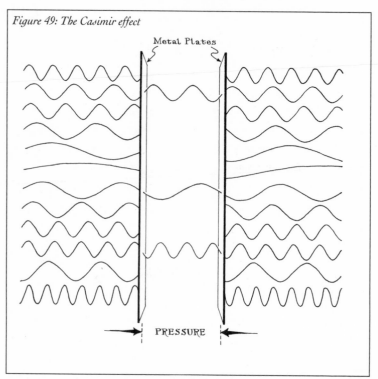

Figure 49: The Casimir effect

Metal Plates

PRESSURE

plates than on the inside. The undiminished zoo of particles presses on the outside of the plates, and without the full complement on the inside, the plates are crushed together, even in the deepest vacuum. This is the force of the vacuum, a force produced by nothing at all. This is the Casimir effect.

Though the Casimir force—a mysterious, phantom force exerted by nothing at all—seems like science fiction, it exists. It is a tiny force and very difficult to measure, but in 1995 the physicist Steven Lamoreaux measured the Casimir effect directly. By putting two gold-covered plates on a sensitive twist-measuring device, he determined how much force it took to counteract the Casimir force between them. The answer—about the weight of one slice of an ant that's been chopped into 30,000 pieces—agreed with Casimir's theory. Lamoreaux had measured the force exerted by empty space.

The Relativistic Zero: The Black Hole

[The star,] like the Cheshire cat, fades from view. One leaves behind only its grin, the other, only its gravitational attraction.

— JOHN WHEELER

Zero in quantum mechanics invests the vacuum with infinite energy. A zero in the other great modern theory — relativity — creates another paradox: the infinite nothing of the black hole.

Like quantum mechanics, the theory of relativity was born in light; this time it was the speed of light that caused the trouble. Most objects in the universe don't have a speed that every observer can agree on. For instance, imagine a small boy who is throwing stones in all directions. For an observer approaching the boy, the stones seem to be going faster than for an observer who is running away; the velocity of the stones seems to depend on your direction and speed. In the same way, the speed of light should depend on whether you are running toward or running away from the lightbulb that's shining on you. In 1887 the American physicists Albert Michelson and Edward Morley tried to measure this effect. They were baffled when they found no difference; the speed of light was the same in every direction. How could this be?

Again, it was the young Einstein who had the answer in 1905. And again, very simple assumptions would have enormous consequences.

The first assumption Einstein made seems fairly obvious. Einstein stated that if a number of people watch the same phenomenon — say, the flight of a raven toward a tree, the laws of physics are the same for each observer. If you compare the notes of a person on the ground and a person on a train moving parallel to the raven, they would disagree about the speed of the raven and the tree. But the eventual outcome of the flight is the same: after a few seconds, the raven arrives at the tree. Both observers agree on the final result, though they might disagree about some of the details. This is the principle of relativity. (In the special theory of relativity, which we are discussing here, there are restrictions on the kind of motion that is allowed. Each observer must be moving with constant velocity in a straight line. In other words, they can't feel an acceleration. With the general theory of relativity, the restrictions are removed.)

The second assumption is a little more troubling, especially since it seems to contradict the principle of relativity. Einstein assumed that everybody—no matter at what speed they are traveling—agrees about the speed of light in a vacuum: about 300 million meters per second, a constant denoted by the letter c. If someone shines a flashlight at you, the light rushes at you at a speed of c. It doesn't matter whether the person holding the flashlight is standing still, running toward you, or running away; the beam of light always travels at a speed of c from your point of view—and everybody else's.

This assumption challenged everything physicists had assumed about the motion of objects. If the raven were acting like a photon, then an observer on the train and the person standing still would have to agree on the value for the raven's speed. That would mean that the two observers would disagree about *when* the raven meets the tree (Figure 50). Einstein realized that there is one way around this: the flow of time changes, depending on an observer's speed. The clock on the train must tick more slowly than the stationary clock. Ten seconds for the observer on the ground might seem like only five seconds for somebody on the train. It's the same thing for a person who zooms away at great speed. Every tick of his stopwatch takes more than a second from a stationary observer's point of view. If an astronaut took a 20-year journey (according to his pocket watch) at nine-tenths of the speed of light, he would come back to Earth having aged 20 years, as expected. But everyone who stayed behind would have aged 46 years.

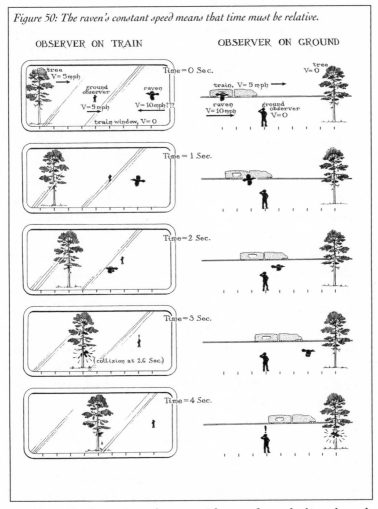

Figure 50: The raven's constant speed means that time must be relative.

Not only does time change with speed, so do length and mass. As objects speed up, they get shorter and heavier. At nine-tenths of the speed of light, for instance, a yardstick would only be 0.44 yards long, and a one-pound bag of sugar would weigh nearly 2.3 pounds — from a stationary observer's point of view. (Of course, this doesn't mean that you would be able to bake more cookies with the same bag of sugar. From the bag's point of view, its weight stays the same.)

This variability in the flow of time might be hard to believe, but it has been observed. When a subatomic particle travels very fast, it survives longer than expected before it decays, because its clock is slow. Also, a very precise clock has been observed to slow down ever so slightly when flown in an airplane at great speed. Einstein's theory works. There was a potential problem though: zero.

When a spaceship approaches the speed of light, time slows down more and more and more. If the ship were to travel at the speed of light, every tick of the clock on board would equal infinite seconds on the ground. In less than a fraction of a second, billions and billions of years would pass; the universe would have already met its ultimate fate and burned itself out. For an astronaut aboard the spaceship, time stops. The flow of time is multiplied by zero.

Luckily, it is not so easy to stop time. As the spaceship goes ever faster, time slows down more and more, but at the same time, the spaceship's mass gets greater and greater. It is like pushing a baby carriage where the baby grows and grows. Pretty soon you are pushing a sumo wrestler—not so easy. If you manage to push the carriage even faster, the baby becomes as massive as a car . . . and then a battleship . . . and then a planet . . . and then a star . . . and then a galaxy. As the baby gets more massive, your push has less and less effect. In the same way, you can take a spaceship and accelerate it, getting it closer and closer to the speed of light. But after a while, it gets too massive to push any longer. The spaceship—or for that matter any other object with mass—never quite reaches the speed of light. The speed of light is the ultimate speed limit; you cannot reach it, much less exceed it. Nature has defended itself from an unruly zero.

However, zero is too powerful even for nature. When Einstein extended the theory of relativity to include gravity, he did not suspect that his new equations—the general theory of relativity—would describe the ultimate zero and the worst infinity of them all: the black hole.

Einstein's equations treat time and space as different aspects of the same thing. We are already used to the idea that if you accelerate, you change the way you move through space; you can speed up or slow down. What Einstein's equations showed was that just as acceleration changes the way you move through space, it changes the way you move through time. It can speed up the way time flows or slow it down. Thus, when you accelerate an object—when you subject it to any force, be it gravity or be it the push of a gigantic cosmic elephant—you change its motion through space and through time: through *space-time.*

It's a difficult concept to grasp, but the easiest way to approach space-time is through an analogy: space and time are like a gigantic rubber sheet. Planets, stars, and everything else sit on that sheet, distorting it slightly. That distortion—the curvature caused by objects sitting on the sheet—is gravity. The more massive the object that is sitting on the sheet, the more the sheet gets distorted, and the larger the dimple around that object. The pull of gravity is just like the tendency of objects to roll into the dimple.

The curvature of the rubber sheet is not only a curvature of space, but a curvature of time as well. Just as space gets distorted close to a massive object, time does, too. It gets slower and slower as the curvature gets greater and greater. The same thing happens with mass. As you get into greatly curved regions of space, bodies' masses effectively increase, a phenomenon known as *mass inflation.*

This analogy explains the orbits of the planets; Earth is simply rolling around in the dimple that the sun makes in the rubber sheet. Light doesn't go in a straight line, but in a curved path around stars—an effect that the British astronomer Sir Arthur Eddington went on an expedition in 1919 to observe. Eddington measured the position of a star during a solar eclipse and spotted the curvature that Einstein had predicted (Figure 51).

Einstein's equations also predicted something much more sinister: the black hole, a star so dense that nothing can escape its grasp, not even light.

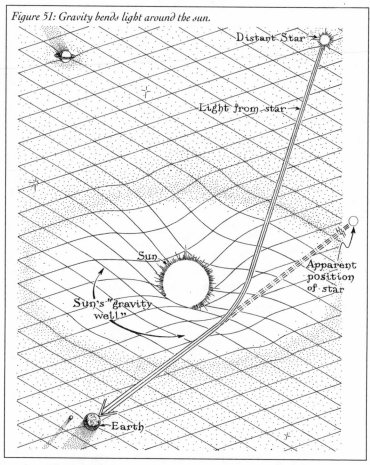

Figure 51: Gravity bends light around the sun.

A black hole begins, like all stars, as a big ball of hot gas—mostly hydrogen. If left to its own devices, a sufficiently large ball of gas would collapse under the weight of its own gravity; it would crush itself into a tiny lump. Luckily for us, stars don't collapse because there is another force at work: nuclear fusion. As a cloud of gas collapses, it gets hotter and denser, and hydrogen atoms slam into one another with increasing force. Eventually, the star gets so hot and dense that the hydrogen atoms stick to one another and fuse, creating helium and releasing large quantities of energy. This energy shoots out from the center of the star, causing it to expand a little bit. During most of its life, a star is in an uneasy equilibrium: the propensity to collapse under its own gravity is balanced by the energy that comes from the fusing hydrogen in its center.

This equilibrium cannot last forever; the star has only a limited amount of hydrogen fuel to burn. After a while, the fusion reaction dims, and the equilibrium is upset. (How long this process takes depends on how big the star is. Ironically, the bigger the star—the more hydrogen it has—the shorter its life, because it burns much more violently. The sun has about five billion years of fuel left, but don't let that make you complacent. The sun's temperature will increase gradually before that, boiling off the oceans and turning Earth into an uninhabitable desert like Venus. We should count ourselves lucky if we have a mere billion years left of life on Earth.) After a drawn-out series of death throes—the precise sequence of events depends, again, on the mass of the star—the star's fusion engine fails, and the star begins to collapse under its own gravity.

A quantum-mechanical law called the Pauli exclusion principle keeps matter from squishing itself into a point. Discovered in the mid-1920s by German physicist Wolfgang Pauli, the exclusion principle states, roughly, that no two things can be in the same place at the same time. In particular, no two electrons of the same quantum state can be forced into the same spot. In 1933, the Indian physicist Subrahmanyan Chandrasekhar realized that the Pauli exclusion principle had only a limited ability to fight against the squeeze of gravity.

As pressure in the star increases, the exclusion principle states that electrons inside must move faster and faster to avoid one another. But there's a speed limit: electrons cannot move faster than the speed of light, so if you put enough pressure on a lump of matter, electrons cannot move fast enough to stop the matter from collapsing. Chandrasekhar showed that a collapsing star that has about 1.4 times the mass of our sun will have enough gravity to overwhelm the Pauli exclusion principle. Above this *Chandrasekhar limit* a star's gravity will pull on itself so strongly that electrons can't stop its collapse. The force of gravity is so great that the star's electrons give up their struggle once and for all; the electrons smash into the star's protons, creating neutrons. The massive star winds up being a gigantic ball of neutrons: a neutron star.

Further calculations showed that when collapsing stars are a little more massive than the Chandrasekhar limit, the pressure of the resulting neutrons — similar to the pressure of electrons — can stave off collapse for a little while; this is what happens in a neutron star. At this point, the star is so dense that every teaspoon weighs hundreds of millions of tons. There is a limit, though, to even the pressure that neutrons can bear. Some astrophysicists believe that a little more squeezing makes the neutrons break down into their component quarks, creating a quark star. But that is the last stronghold. After that, all hell breaks loose.

When an extremely massive star collapses, it disappears. The gravitational attraction is so great that physicists know of no force in the universe that can stop its collapse — not the repulsion of its electrons, not the pressure of neutron against neutron or quark against quark — nothing. The dying star gets smaller and smaller and smaller. Then . . . zero. The star crams itself into zero space. This is a black hole, an object so paradoxical that some scientists believe that black holes can be used to travel faster than light — and backward in time.

The key to a black hole's strange properties is the way it curves space-time. A black hole takes up no space at all, but it still has mass. Since the black hole has mass, it causes space-time to curve. Normally, this would not cause a problem. As you approach a heavy star, the curvature gets greater and greater, but once you have passed the outer edge of the star itself, the curvature decreases again, bottoming out at the center of the star. In contrast, a black hole is a point. It takes up zero space, so there is no outer edge, no place where space begins to flatten out again. The curvature of space gets greater and greater as you approach a black hole, and it never bottoms out. The curvature goes off to infinity because the black hole takes up zero space; the star has torn a hole in space-time (Figure 52). The zero of a black hole is a singularity, an open wound in the fabric of the universe.

This is a very troublesome concept. The smooth, continuous fabric of space-time might have tears in it, and nobody knows quite what happens in the region of those tears. Einstein was so disturbed by the idea of singularities that he de-

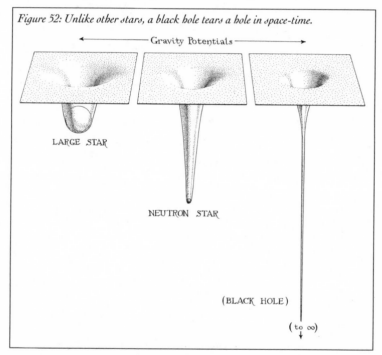

Figure 52: Unlike other stars, a black hole tears a hole in space-time.

Gravity Potentials

LARGE STAR

NEUTRON STAR

(BLACK HOLE)

(to ∞)

nied the existence of black holes. He was wrong; black holes do exist. However, the singularity of a black hole is so ugly, so dangerous, that nature tries to shield it, preventing anyone from seeing the zero at the center of a black hole and returning to tell the tale. Nature has a "cosmic censor."

The censor is gravity itself. If you toss a rock upward, it will curve back down, pulled back by the earth's gravity. But if you throw a rock fast enough, it won't curve back down to earth; it will zoom out of the earth's atmosphere and escape the earth's gravitational pull. This is roughly what NASA does when it sends a spacecraft to Mars. The minimum speed you need to throw the rock to enable it to escape is called, naturally enough, the *escape velocity*. Black holes are so dense that if you get too close—past the so-called event horizon— the escape velocity is faster than the speed of light. Past the event horizon the pull of a black hole's gravity is so strong— and space is so curved—that nothing can escape, not even light.

Even though a black hole is a star, none of the light it shines ever escapes past the event horizon; that's why it's black. The only way to view a black hole's singularity is to go beyond the event horizon and see for yourself. However, even if you had an impossibly strong spacesuit that kept you from being stretched into a piece of astronaut spaghetti, you could never tell anybody about what you saw. Once you pass the event horizon, signals you broadcast can't escape the black hole's pull—neither can you. Traveling beyond the edge of the event horizon is like stepping off the edge of the universe. You will never return. This is the power of the cosmic censor.

Even though nature tries to shield the singularities of black holes, scientists know that black holes exist. In the direction of the constellation Sagittarius, at the very center of our galaxy, sits a supermassive black hole that weighs as much as two-and-a-half million suns. Astronomers have watched stars dance around an invisible partner; the stars' motions reveal the presence of the black hole even if the black hole is not visible. However, though scientists can detect black holes, they still haven't spotted the zeros at their centers, since the ugly singularities are shielded by the event horizon.

This is a good thing. If there were no event horizon, no cosmic censor that shields the singularity from the rest of the universe, very strange things might happen. In theory, a *naked singularity* with no event horizon might allow you to travel faster than light or backward in time. This could be done with a structure known as a *wormhole.*

Back in the rubber-sheet analogy, a singularity is a point of infinite curvature; it is a hole in the fabric of space and time. Under certain circumstances that hole can be stretched out. For instance, if a black hole is spinning or has an electric charge, mathematicians have calculated that the singularity is not a point—a pinpoint hole in space-time—but a ring. Physicists have speculated that two of these stretched-out singularities might be linked with a tunnel: a wormhole (Figure 53). A person who travels through a wormhole will emerge at another point in space—and perhaps in time. Just as wormholes can, in theory, send you halfway across the universe in the blink of an eye, they can send you backward and forward in time (see appendix E). You might even be able to track down your mother and kill her before she meets your father, preventing you from being born and causing a terrible paradox.

A wormhole is a paradox caused by a zero in the equations of general relativity. Nobody truly knows whether or not wormholes exist—but NASA is hoping that they do.

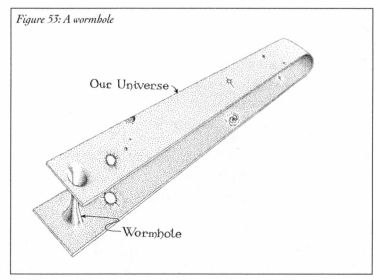

Figure 53: A wormhole

Our Universe

Wormhole

Something for Nothing?

There's no such thing as a free lunch.
—"THE SECOND LAW OF THERMODYNAMICS"

NASA hopes that zero might hold the secret to traveling to distant stars. In 1998, NASA held a symposium entitled Physics for the Third Millennium, where scientists debated the merits of wormholes, warp drives, vacuum-energy engines, and other far-out ideas.

The problem with space travel is that there is nothing to push against. When you swim through a pool, you push against the water, forcing it backward and pushing you forward. When you walk on the ground, your feet are pushing against the floor, providing the force to drive you forward. In space, there is nothing to push backward; you can paddle all you want, but you'll get nowhere.

Rockets bring their own supply of stuff to push against. Rocket fuel burns in the engine and is sent out the back of the rocket, driving the spaceship forward, just as the rush of air out of a balloon sends it flying around the room. But tossing away fuel is an expensive and cumbersome way to go places, and even modern improvements on the chemical engine, such as engines that use electricity to throw stuff out the back of a rocket, are unable to provide the fuel efficiency to send probes to distant stars in a reasonable amount of time. To get even to the nearest star, you would need an enormous amount of fuel to jettison out the back of the rocket—a tremendous waste.

Physicist Marc Millis heads NASA's Breakthrough Propulsion Project and hopes to overcome this problem with the physics of zero. Unfortunately, the zeros of black holes— singularities—look like unlikely candidates in the short term. Not only is it extremely difficult to create a naked singularity that a wormhole needs, but it also appears that even a naked singularity will tear space travelers to shreds. In 1998 two physicists from the Hebrew University in Jerusalem showed that even a spinning or charged black hole — with a nice, ring-shaped singularity — will kill an astronaut, thanks to mass inflation. As you fall toward the singularity, the black hole's mass appears to grow and grow to infinity. The gravitational tug is so strong that you'd be torn apart in a fraction of a second. Wormholes would be hazardous to your health.

Even if the zeros at the center of black holes don't provide an easy way to travel through space, the zero of quantum mechanics provides an alternative: the zero-point energy might be the ultimate fuel. It is here that the mainstream of physics ends and the fringe begins.

According to Millis, astronauts might harness the energy in the vacuum to push a spaceship, just as mariners harnessed the wind to drive a frigate. "I'm making an analogy to the Casimir effect, where you can push plates together with a noticeable radiation pressure from the vacuum," he says. "If there were any way to get asymmetric forces out of that, where you get force in one direction and not the other, you'd get a propulsive force." Unfortunately, so far the Casimir effect seems to be symmetric; both plates collapse and pull each other together. The action of one has an equal and opposite reaction on the other. But if there were some sort of quantum sail, a one-way mirror that reflected virtual particles on one side but let them pass unhindered through the other, the vacuum energy would push the whole object toward the unreflective half of the sail. Millis admits that nobody has any clue how to do this. "There are no theories how to engineer the device," he says sadly.

The problem is that the laws of physics say you can't get something for nothing; just as the frigate lowers the speed of the wind, the quantum sail would have to lower the energy of the vacuum. How can you modify nothing?

Harold Puthoff, the director of the Institute for Advanced Studies in Austin, Texas, believes that a quantum sail would simply alter the properties of the vacuum. (Puthoff is best known for his 1974 paper in *Nature* that purported to prove that Uri Geller and other psychics could view objects remotely—without their eyes. This conclusion was not in the mainstream of science.) "The vacuum decays to a slightly lower state," says Puthoff. If so, then quantum sails are just the beginning; it would be possible to make engines that run solely on zero-point energy. Their only drawback would be that the fabric of the universe would fall apart. Slowly. "We'd never make a dent. It's like scooping up cupfuls of water from the ocean," says Puthoff.

It might also destroy the universe.

There is no question that the vacuum has energy; the Casimir force is witness to that fact. But is it possible that the energy of the vacuum is truly the lowest possible energy? If not, danger might be lurking in the vacuum. In 1983 two scientists suggested in *Nature* that tinkering with the energy of the vacuum might cause the universe to self-destruct. The paper argued that our vacuum might be a "false" vacuum in an unnaturally energetic state—like a ball perched precariously on the side of a hill. If we give the vacuum a big enough nudge, it might start rolling down the hill—settling into a lower energy state—and we would not be able to stop it. We would release a huge bubble of energy that expands at the speed of light, leaving a vast trail of destruction in its wake. It might be so bad that every one of our atoms would be torn apart during the apocalypse.

Luckily, this is an extremely unlikely scenario. Our universe has lasted billions of years, and it's improbable that we are living in such a precarious state; cosmic-ray collisions would probably already have "sparked" the vacuum with enough energy to cause such a disaster were it possible. This hasn't stopped some believers—physicists included—from picketing high-energy laboratories like Fermilab; they believe that a high-energy collision could cause a spontaneous collapse of the vacuum. Even if those concerns were valid, it seems all but impossible to propel a spaceship with zero-point energy. However, Puthoff believes he has a way to extract energy from the void.

In theory, scientists can get energy from the Casimir effect even at absolute zero in the bleakest part of the vacuum of space. Two plates generate heat when they smack together — heat that can be converted to electricity. Alas, the plates have to be pried apart again, which requires more energy than was initially produced; most scientists believe that this fact kills the idea of making a perpetual-motion machine that runs on vacuum energy. But Puthoff thinks he sees several ways around this dilemma. One is to use plasmas instead of plates.

A plasma, a gas of charged particles, is just like a metal plate as far as the Casimir effect is concerned. A conducting, cylinder-shaped gas would be compressed by the zero-point fluctuations just as plates are forced together. The collapse would heat the plasma, releasing energy. Unlike metal plates, plasmas could be made easily with a bolt of electricity, according to Puthoff, and instead of having to pry the plates apart again, the plasma "ash" is discarded. Puthoff gingerly claims to have gotten out 30 times more energy with this method than was put in. "There's some evidence; we've even got a patent," he says. However, Puthoff's device is one in a long line of "free energy" machines — none of which, in the past, have withstood scientific scrutiny. It is unlikely that his device to harness the zero-point energy will be any different.

According to quantum mechanics and general relativity, the power of zero is infinite, so it's no surprise that people are hoping to tap its potential. But for the time being, it appears that nothing will come of nothing.

Chapter **8**
Zero Hour at Ground Zero

[ZERO AT THE EDGE OF SPACE AND TIME]

Alien they seemed to be:

No mortal eye could see

The intimate welding of their later history . . .

— THOMAS HARDY,
"THE CONVERGENCE OF THE TWAIN"

M odern physics is a struggle of two titans. General relativity holds sway in the realm of the very, very big: the most massive objects in the universe, such as stars, solar systems, and galaxies. Quantum mechanics rules the domain of the very, very small: atoms and electrons and subatomic particles. It would seem that these two theories could live in harmony together, each dictating the rules of physics for different aspects of the universe.

Unfortunately, there are objects that lie in both realms. Black holes are very, very massive, so they are subject to the laws of relativity; at the same time, black holes are very, very tiny, so they are in the domain of quantum mechanics. And far from agreeing, the two sets of laws clash at the center of a black hole.

Zero dwells at the juxtaposition of quantum mechanics and relativity; zero lives where the two theories meet, and zero causes the two theories to clash. A black hole is a zero in the equations of general relativity; the energy of the vacuum is a zero in the mathematics of quantum theory. The big bang, the most puzzling event in the history of the universe, is a zero in both theories. The universe came from nothing—and both theories break down when they try to explain the origin of the cosmos.

To understand the big bang, physicists must marry quantum theory with relativity. In the past few years they have begun to succeed, creating a monster theory that explains the quantum-mechanical nature of gravity, allowing them to peer at the very creation of our universe. All they had to do was banish zero.

The Theory of Everything is, in truth, a theory of nothing.

Zero Banished: String Theory

The problem is, when we try to calculate all the way down to zero distance, the equation blows up in our face and gives us meaningless answers—things like infinity. This caused a lot of trouble when the theory of quantum electrodynamics first came out. People were getting infinity for every problem they tried to calculate!

— RICHARD FEYNMAN

General relativity and quantum mechanics were bound to be incompatible. The universe of general relativity is a smooth rubber sheet. It is continuous and flowing, never sharp, never pointy. Quantum mechanics, on the other hand, describes a jerky and discontinuous universe. What the two theories have in common — and what they clash over — is zero.

The infinite zero of a black hole — mass crammed into zero space, curving space infinitely — punches a hole in the smooth rubber sheet. The equations of general relativity cannot deal with the sharpness of zero. In a black hole, space and time are meaningless.

Quantum mechanics has a similar problem, a problem related to the zero-point energy. The laws of quantum mechanics treat particles such as the electron as points; that is, they take up no space at all. The electron is a *zero-dimensional* object, and its very zerolike nature ensures that scientists don't even know the electron's mass or charge.

This seems like a silly statement. It has been nearly a century since scientists measured the electron's mass and charge. How could physicists not know something that has been measured? The answer lies with zero.

The electron that scientists see in the laboratory — the electron that physicists, chemists, and engineers have known and loved for decades — is an impostor. It is not the true electron. The true electron is hidden in a shroud of particles, made up of the zero-point fluctuations, those particles that constantly pop in and out of existence. As an electron sits in the vacuum, it occasionally absorbs or spits out one of these particles, such as a photon. The swarm of particles makes it difficult to get a measurement of the electron's mass and charge, because the particles interfere with the measurement, masking the electron's true properties. The "true" electron is a bit heavier and carries a greater charge than the electron that physicists observe.

Scientists might get a better idea of the true mass and charge of the electron if they could get a little closer; if they could invent a tiny device that could get a short distance inside the cloud of particles, they would be able to see more clearly. According to quantum theory, as the measuring device gets past the first few virtual particles on the rim of the cloud, scientists would see the mass and charge of the electron go up, and as the probe gets closer and closer to the electron, it would pass more and more virtual particles, so the observed mass and charge go up and up. As the probe approaches zero distance from the electron, the number of particles it passes goes up to infinity—so the probe's measurements of the mass and charge of the electron also go to infinity. According to the rules of quantum mechanics, the zero-dimensional electron has infinite mass and infinite charge.

As with the zero-point energy, scientists learned to ignore the infinite mass and charge of the electron. They don't go all the way to zero distance from the electron when they calculate the electron's true mass and charge; they stop short of zero at an arbitrary distance. Once a scientist chooses a suitably close distance, all the calculations using the "true" mass and charge agree with one another. This is a process called renormalization. "It is what I would call a dippy process," wrote physicist Richard Feynman, even though Feynman won his Nobel Prize for perfecting the art of renormalization.

Just as zero punches a hole in the smooth sheet of general relativity, zero smooths and spreads out the sharp point charge of the electron, covering it in a fog. However, since quantum mechanics deals with zero-dimensional particle-points such as the electron, technically all particle-particle interactions in quantum theory deal with infinities: they are singularities. When two particles merge, for instance, they meet at a point: a zero-dimensional singularity. This singularity makes no sense in quantum mechanics or in general relativity. Zero is the wrench in the works of both great theories. So physicists simply got rid of it.

It is not obvious how to get rid of zero, as zero appears and reappears throughout time and space. Black holes are zero-dimensional, as are particles such as the electron. Electrons and black holes are real things; physicists can't simply will them away. But scientists can give black holes and electrons an extra dimension.

This is the reason for *string theory*, which was created in the 1970s when physicists began to see the advantages of treating every particle as a vibrating string rather than as a dot. If electrons (and black holes) are treated as one-dimensional, like a loop of string, instead of as zero-dimensional, like a point, the infinities in general relativity and quantum mechanics miraculously disappear. For instance, the renormalization trouble—the infinite mass and charge of the electron—vanishes. A zero-dimensional electron has an infinite mass and charge because it is a singularity; as you get closer and closer to it, your measurements zoom off to infinity. However, if the electron is a loop of string, the particle is no longer a singularity. This means that the mass and charge don't go off to infinity, because you are no longer passing an infinite cloud of particles as you approach the electron. Furthermore, as two particles merge, no longer do they meet at a pointlike singularity; they form a nice, smooth, continuous surface in space-time (Figures 54, 55).

In string theory different particles are really the same type of

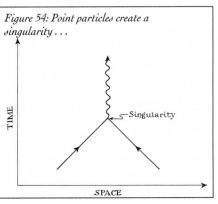

Figure 54: *Point particles create a singularity . . .*

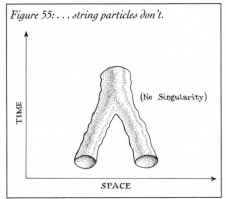

Figure 55: *. . . string particles don't.*

string, just wiggling in different ways. Everything in the universe is made up of these strings, which are about 10^{-33} centimeters across; comparing the size of a string to the size of a neutron is like comparing the size of a neutron to the size of our solar system. From the perspective of beings as large as we are, the loops look like points because they are so tiny. Distances (and times) smaller than the size of the loops no longer matter; they don't make any physical sense. In string theory, zero has been banished from the universe; there is no such thing as zero distance or zero time. This solves all the infinity problems of quantum mechanics.

Banishing zero also solves the infinity problems in general relativity. If you imagine a black hole as a string, no longer do objects fall through a rip in the fabric of space-time. Instead, a particle loop approaching a black-hole loop stretches out and touches the black hole. The two loops tremble, tear, and form one loop: a slightly more massive black hole. (Some theorists believe that the act of merging a particle to a black hole creates bizarre particles such as *tachyons:* particles with imaginary mass that travel backward in time and move faster than light. Such particles might be admissible in certain versions of string theory.)

Removing zero from the universe might seem like a drastic step, but strings are much more tractable than dots; by eliminating zero, string theory smooths out the discontinuous, particle-like nature of quantum mechanics and mends the gashes torn in general relativity by black holes. With these problems patched over, the two theories are no longer incompatible. Physicists began to think that string theory would unify quantum mechanics with relativity; they believed that it would lead to the theory of quantum gravity — the Theory of Everything that explains every phenomenon in the universe. However, string theory had some problems. For one thing, it required 10 dimensions to work.

For most people, four dimensions are one too many. It is easy to see three of them: left-right, front-back, and up-down represent the three directions we can move in. The fourth arrived when Einstein showed that time was similar to these three dimensions; we are constantly moving through time like a car that's speeding down a highway. The theory of relativity shows that just as we can change how quickly we rush down a highway, we can change the rate at which we move through time—the faster we go through space, the quicker we move through time. To understand Einstein's universe, we have to accept the idea that time is the fourth dimension.

Four is reasonable—but 10? We can measure four dimensions, but what happened to the other six dimensions? According to string theory, they are rolled up like little balls, too tiny to see. When you pick up a piece of paper, it seems two-dimensional. It has length and breadth, but it doesn't seem to have any depth at all. Nevertheless, if you take a magnifying glass and gaze at the edge of the piece of paper, you begin to see that it has a wee bit of depth. You need a tool to help you see it, but that third dimension is there, too tiny to see under normal conditions. The same is true with those extra six dimensions. In everyday life they are way too tiny to see; they are too small to detect even with the most powerful equipment that we could possibly manufacture in the near future.

What do these six extra dimensions *mean*? Nothing, really. They don't measure anything that we are accustomed to, like length, breadth, width, or time. They are simply mathematical constructs that make the mathematical operations in string theory work in the manner that they have to. Like imaginary numbers, we can't see them or feel them or smell them, even though they are necessary for doing calculations. Though it is a strange concept physically, it is the predictive power of the equations that interests scientists, rather than their comprehensibility—and an extra six dimensions do not constitute an insurmountable problem, mathematically. Spotting them might. (Ten seems small nowadays. In the past few years physicists realized that the many competing varieties of string theory are actually, in a sense, the same thing. Scientists realize now that these theories are dual to each other just as Poncelet realized that lines and points were dual to each other. Scientists now believe that there is a monster theory that underlies all of these competing theories: the so-called M-theory, which lives in 11 dimensions, not 10.)

Strings (or their more general counterparts, *branes*, a term for multidimensional membranes) are so tiny that no instrument can hope to spot them—at least until our civilization becomes much more advanced. Particle physicists look at the subatomic realm with particle accelerators: they use magnetic fields or other means to get tiny particles moving very fast; when these particles collide with one another, they spit off fragments. Particle accelerators are the microscopes of the subatomic world, and the more energy you put into those particles—the more powerful the microscope—the smaller the objects you can see.

The Superconducting Super Collider, a multibillion dollar project that was contemplated until the early 1990s, was going to be the most powerful particle accelerator ever built. It was to have more than 10,000 magnets arranged in a loop 54 miles around, about the size of the beltway highway that circles Washington, D.C. This is still not nearly powerful enough to see strings or curled-up dimensions—viewing strings would require a particle accelerator about 6,000,000,000,000,000 miles around. Even traveling at the speed of light, a particle would take 1,000 years to make the circuit.

No instrument currently imaginable will give scientists the power to observe strings directly; nobody can think of an experiment that will give physicists evidence about whether black holes and particles are, indeed, strings. This is the chief objection to string theory. Because science is based upon observation and experiment, some critics argue that string theory is not science but philosophy. (A recent set of theories proposes that some of these rolled-up dimensions might be 10^{-19} centimeters or even larger, which would put them within the realm of experimentation. But at the moment, these theories are considered rogues—interesting ideas, but very long shots at best.)

Newton's laws of motion and gravitation gave physicists an explanation for the way planets and objects move through the universe. Whenever a new comet was discovered, it gave additional support to Newton's calculations. There were a few problems. Mercury's orbit, for instance, wobbled in a way that disagreed with what Newton predicted, but on the whole, Newton's theories were tested again and again, and they usually passed.

Einstein's theories corrected Newton's errors; they explained Mercury's wobble, for instance. These theories also made testable predictions about the way gravity works. Eddington observed the bending of starlight during a solar eclipse, confirming one of those predictions.

String theory, on the other hand, ties together a number of existing theories in a very pretty way, and makes a number of predictions about the way black holes and particles behave, but none of those predictions are testable or observable. While string theory might be mathematically consistent, and even beautiful, it is not yet science.[*]

For the foreseeable future, banishing zero from the universe with string theory is a philosophical idea rather than a scientific one. String theory might well be correct, but we may never have the means to find out. Zero has not yet been banished; indeed, zero seems to be what created the cosmos.

---

[*]Yes, mathematics can be "beautiful" or "ugly." Just as it's hard to describe what makes a piece of music or a painting aesthetically pleasing, it's equally difficult to describe what makes a mathematical theorem or a physical theory beautiful. A beautiful theory will be simple, compact, and spare; it will give a sense of completeness and often an eerie sense of symmetry. Einstein's theories are particularly beautiful, as are Maxwell's equations. But for many mathematicians, an equation discovered by Euler, $e^{i\pi} + 1 = 0$, is the paragon of mathematical beauty, because this extremely simple, compact formula relates all the most important numbers in mathematics in a totally unexpected way.

The Zeroth Hour: The Big Bang

Hubble's observations suggested that there was a time, called the big bang, when the universe was infinitesimally small and infinitely dense. Under such conditions all the laws of science, and therefore all ability to predict the future, would break down.

—STEPHEN HAWKING, *A BRIEF HISTORY OF TIME*

The universe was born in zero.

Out of the void, out of nothing at all, came a cataclysmic explosion that created all the matter and energy that the entire universe is made of. This event—the big bang—was a horrible idea to many scientists and philosophers. It took a long time before astrophysicists came to agree that our universe was finite—that it did, in fact, have a beginning.

The prejudice against a finite universe is ancient. Aristotle rejected the creation of the universe out of the void because he believed that the void could never exist. But this caused a paradox. If the universe could not spring forth from the void, then *something* had to be floating about before the birth of the universe; there had to be a universe before the universe was born. To Aristotle, the only possible way out of this quandary was to assume that the universe was eternal. It had always existed in the past, and would always exist.

Western civilization eventually had to make a choice between Aristotle and the Bible, which says that the finite universe sprang forth from the void and prophesies its ultimate destruction. Though the Semitic biblical cosmos toppled the Aristotelian one, the idea of an eternal, unchanging universe was not expunged completely, enduring even to the twentieth century. It led Einstein to what he called the greatest mistake of his career.

To Einstein, the general theory of relativity had a crucial flaw. It foretold the end of the universe. According to the equations of general relativity, the universe was unstable. There were only two choices, and both were equally unpleasant.

One possibility was that the universe would collapse under its own gravity. As the universe gets smaller and smaller, it heats up more and more. It burns brightly with radiation, destroying all life and eventually destroying the atoms that make up matter. It would be death by fire. Eventually, the universe would crunch itself into a zero-dimensional point—like a black hole—and would disappear forever.

The other possibility is, if anything, more grim. The universe would expand forever. Galaxies would become ever more distant from one another, and the star stuff that drives all the energetic reactions in the universe would become more rarefied. Stars would burn out as they exhausted their fuel, and galaxies would become darker and darker—and then cold and silent. The cold, dead matter of the stars would decay away, leaving nothing but a smear of radiation that spreads equally throughout the universe. The cosmos would be a cold soup of dimming light. It would be death by ice.

To Einstein, these ideas were abhorrent. Like Aristotle, he implicitly assumed that the universe was static, constant, and eternal. The only way out was to "correct" his equations of general relativity to stave off the impending destruction. He did this by adding a *cosmological constant,* an as yet undetected force that counteracts the force of gravity. The cosmological constant's push would balance out gravity's pull; instead of collapsing, the universe could stay in a steady balance, neither collapsing nor expanding. Postulating the existence of such a mysterious force was a desperate act. "I have . . . again perpetrated something about gravitation theory which somewhat exposes me to the danger of being confined in a madhouse," wrote Einstein, but he was so worried about the impending destruction of the universe that he was forced to take such a dramatic step.

Einstein wasn't bundled off to an asylum. Einstein had proposed stranger things and had been entirely right. However, this time he was not so lucky. The stars themselves destroyed Einstein's vision of a static, eternal cosmos.

In 1900 the Milky Way was the known universe. Astronomers had little idea that anything lay beyond our own dusty little disk of stars. Though astronomers had spotted some glowing, swirly clouds, there was little reason to believe that they were anything but glowing gas inside our galaxy. In the 1920s that all changed, thanks to an American astronomer named Edwin Hubble.

A special type of star, called a Cepheid variable, had a property that allowed Hubble to measure the distance to faraway objects. Cepheid stars pulsate, getting brighter and dimmer in a very predictable way; the way they pulsate is closely related to how much light they put out. They are *standard candles*, objects of known brightness, and became a key tool for Hubble. They were like the headlights of a train.

If you watch a train coming at you, you will see that its headlight gets brighter and brighter as it approaches. If you know how much light the headlight puts out—if the headlight is a standard candle—you can tell how bright the headlight will appear at any given distance. The closer it gets, the brighter it seems. The same logic works in reverse; if you know how much light a train's headlight emits, you can measure its apparent brightness and calculate the train's distance from you.

Hubble did the exact same thing with Cepheid stars. Most stars he saw were tens or hundreds or thousands of light-years away. But when he found a Cepheid blinking in one of these swirly clouds—the Andromeda nebula, as it was then called—he measured the light and calculated that the nebula was a million light-years away, far beyond the outer reaches of our galaxy. Andromeda was not a cloud of glowing gas; it was a cloud of stars so distant that they looked like a smear rather than individual points of light. Other swirly galaxies were even more distant. Today, astronomers suspect that the universe is about 15 billion light-years across and peppered everywhere with clusters of galaxies.

This was an astounding discovery; the universe was millions of times bigger than previously suspected. As amazing as this observation was, it was not what Hubble is best remembered for. Hubble's second discovery was what shattered Einstein's eternal universe.

Hubble measured the distance to galaxy after galaxy with his Cepheid stars, but soon began to notice an alarming pattern: all the galaxies were fleeing with high speed, shooting away from the Milky Way at speeds of hundreds of miles a second or more. The galaxies were so distant that even these great velocities were not directly measurable.

The only way to clock the speed of a galaxy is by using the Doppler effect — the same principle used in state troopers' radar guns. You might have noticed that when a train zooms by, the pitch of its horn changes. As the train approaches, its horn is high-pitched, but as it passes you, all of a sudden its pitch drops dramatically. This happens because the motion of the train crushes the sound waves in front of it (making a higher-frequency, higher-pitch tone) and stretches out the waves behind it (making a lower-frequency, lower-pitch tone) (Figure 56). This is the Doppler effect, and it works with light, too. If a star is moving toward Earth, the light is crushed and has a higher frequency than normal; it is shifted toward the blue end of the spectrum, *blueshifted*. If a star is moving away, the opposite happens; the light is stretched out and *redshifted*.

Police can tell how fast a car is going by testing how light — in the form of radio waves — reflected off the speeding vehicle gets shifted. In the same way, by looking at how a star's light spectrum gets shifted, astronomers can deduce how fast the star is moving — toward us or away.

Hubble combined the distance data with Doppler speed data, and found something shocking. Not only were galaxies speeding away from us in all directions, the farther away the galaxies were, the faster they were going away.

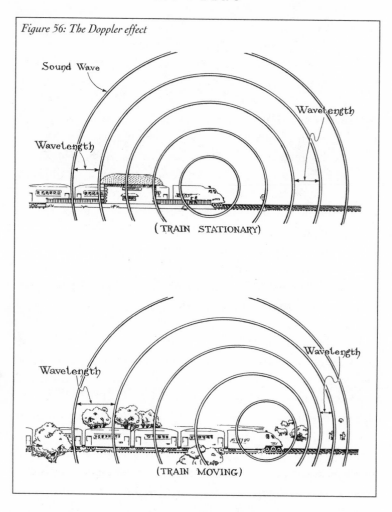

Figure 56: The Doppler effect

Sound Wave

Wavelength

Wavelength

(TRAIN STATIONARY)

Wavelength

Wavelength

(TRAIN MOVING)

How could this be? Imagine a polka-dotted balloon; the polka dots are like galaxies, while the balloon itself is the fabric of space-time. As the balloon inflates, the dots get farther and farther apart from one another. From any one dot's perspective, the other dots are all rushing away, and the more distant dots are rushing faster than the close dots (Figure 57). The universe seemed to be expanding, like a balloon. (The balloon analogy has one flaw. Unlike the polka dots, which also get bigger as the balloon expands, the galaxies

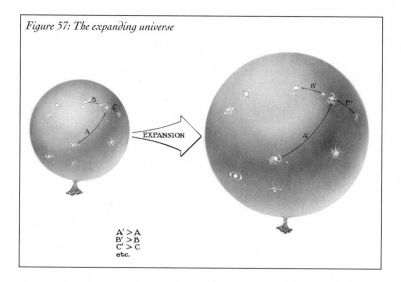

Figure 57: The expanding universe

A' > A
B' > B
C' > C
etc.

are staying about the same size, held together by their own gravity.)

As time runs forward, the universe expands and expands. Looking at it another way, if you had a film of the universe's history and ran the film backward, the universe would shrink and shrink. At some point the balloon would shrivel and wither, getting smaller and smaller, and then eventually disappear as a point—the singularity at the beginning of time and space. This is the primal zero, the birthplace of the universe: the big bang, a furious explosion that created the cosmos. It is from this singularity that all the matter and energy in the universe spewed forth, creating all the galaxies, stars, and planets that have ever—and will ever—exist. The universe had a beginning, about 15 billion years in the past, and space has been expanding ever since. Einstein's hope for a steady, eternal universe was all but dead.

One glimmer of hope remained, one alternative to the big bang: steady-state theory. Some astronomers proposed that there were fountains that spat out matter, and the galaxies moved away from these founts, aging and dying. Though the individual galaxies zoomed away and died, the entire universe as a whole never changed. It was always in equilibrium, constantly replenishing itself. Aristotle's eternal universe still survived.

For a time, big bang theory and steady-state theory lived side by side, alternatives that astronomers chose between depending on their philosophy. In the mid-1960s, that all changed. Steady-state theory was killed by what scientists had mistaken for pigeon droppings.

In 1965 several astrophysicists at Princeton University were calculating what would have happened right after the big bang. The entire universe must have been incredibly hot and dense; it would have been glowing with bright light. That light would not have disappeared as the balloon universe expanded; it would instead have gotten stretched out as the rubber fabric of space-time stretched. A few calculations later, the Princeton physicists realized that this light had to be in the microwave region of the spectrum, and had to be coming from all directions. This *cosmic background radiation* was the afterglow of the big bang. It would give physicists the first direct evidence that the big bang was correct and that steady state was wrong.

The Princeton scientists did not have long to wait before their prediction was confirmed. At Bell Labs in nearby Murray Hill, New Jersey, two engineers had been testing out sensitive microwave-detecting equipment. For all their tinkering, they could not get the equipment to work just right. There was a background hiss of microwave noise—like static on a radio program—that they could not get rid of. At first they thought that pigeons defecating in their antenna were to blame, but after chasing away the birds and cleaning out the droppings, the hiss remained. They tried everything they could think of to get rid of the noise, but nothing worked. Then when the engineers heard of the Princeton group's work, they realized that they had found the cosmic background radiation. The noise wasn't pigeon droppings. It was the scream of light from the big bang, stretched and distorted into a whisper's hiss. (For their discovery the engineers, Arno Penzias and Robert Wilson, got the Nobel Prize. The Princeton physicists, notably Bob Dicke and P. J. E. "Jim" Peebles, got nothing—hardly fair in many scientists' opinions. The Nobel committee tends to reward painstaking and careful experiments more than important theory.)

The big bang had been spotted; the myth of the static universe was dead. As unappealing as the idea of a finite universe was, physicists gradually accepted the big bang and agreed that the universe had a beginning. However, there were still problems with the theory. For one thing, the universe is somewhat lumpy. Knots of dense galaxies are separated by vast voids. At the same time, the universe is not *too* lumpy; it looks roughly the same in all directions, so all the matter did not wind up in one huge glob. If the universe had come from a singularity, with all probability the energy from the big bang should have covered the entire balloon fairly evenly or wound up in one big lump; the balloon should be evenly shaded or it should have one giant spot, rather than being polka-dotted. Something had to account for that just-right amount of lumpiness. More troubling still, where did the singularity of the big bang come from? Zero holds the secret.

The zero of the vacuum might explain the lumpiness of the universe. Since the vacuum everywhere in the universe is seething with a quantum foam of virtual particles, the fabric of the universe is filled with infinite zero-point energy. Under the right conditions this energy is able to push objects around; in the early universe it might have pushed objects apart.

In the 1980s physicists suggested that the zero-point energy in the early universe was greater than it is today. That extra energy would try to expand in all directions, pushing the fabric of space and time outward with great speed. It would inflate the balloon with a huge burst of power, smoothing out the lumpiness of the universe in the same way a breath of air smooths out the wrinkles of a balloon. This explains why the universe is relatively smooth. But the vacuum of the first few moments is a false vacuum; its zero-point energy is unnaturally large. The higher energy state of the zero-point energy makes it inherently unstable, and very quickly—in less than a millionth of a millionth of a millionth of a millionth of a second—the false vacuum would collapse, reverting to the true vacuum, with its everyday zero-point energy that we observe in our universe. It was like a pot of water that was instantly flash heated to a huge temperature. Little bubbles of "true" vacuum would have formed and expanded at the speed of light. Our observable universe sits inside one of these bubbles—or several of them that got linked together. The asymmetry of the universe could be explained by the asymmetrical nature of these expanding bubbles that formed and merged. According to this theory of inflation, it is the nonzero zero-point energy that created the stars and galaxies.

Zero might also hold the secret of what created the cosmos. Just as the nothingness of the vacuum and the zero-point energy spawn particles, they might spawn universes. The froth of quantum foam, the spontaneous birth and death of particles, might explain the origin of the cosmos. Perhaps the universe is just a quantum fluctuation on a grand scale—an enormous singular particle that came into existence out of the ultimate vacuum. This cosmic egg would explode, inflate, and create the space-time of our universe. It may be that our universe is simply one of many fluctuations. Some physicists believe that the singularities at the center of black holes are windows into the primordial quantum foam before the big bang—and the froth of foam at the center of a black hole, where time and space have no meaning, is constantly creating countless numbers of new universes that bubble off, inflate, and create their own stars and galaxies. Zero might hold the secret to our existence—and the existence of an infinite number of other universes.

Zero is so powerful because it unhinges the laws of

physics. It is at the zero hour of the big bang and the ground zero of the black hole that the mathematical equations that describe our world stop making sense. However, zero cannot be ignored. Not only does zero hold the secret to our existence, it will also be responsible for the end of the universe.

Chapter ∞
Zero's Final Victory

[END TIME]

This is the way the world ends

Not with a bang but a whimper.

—T. S. ELIOT, "THE HOLLOW MEN"

While some physicists are trying to eliminate zero from their equations, others are showing that zero may have the last laugh. Even though scientists might never unlock the secrets of the universe's birth, they are on the brink of understanding its death. The ultimate fate of the cosmos lies with zero.

Einstein's gravitational equations didn't allow for a static, unchanging universe. They did, however, allow for several other fates, which depend on the amount of mass in the cosmos. In the case of a light universe, the balloon of space-time could expand forever, getting bigger and bigger. The stars and galaxies would wink out, one by one. The universe grows cold and dies a *heat death*. However, if there is enough mass—

galaxies, galaxy clusters, and unseen dark matter—the initial push given by the big bang wouldn't be enough to allow the balloon to inflate forever. The galaxies would tug on one another, eventually pulling the fabric of space-time together; the balloon would begin to deflate. The deflation would get faster and faster, the universe would get hotter and hotter, and it would eventually end in a backward big bang: the *big crunch*. Which will be our fate: big crunch or heat death? The answer is at hand.

When astronomers peer at a distant galaxy, they are looking backward in time. A nearby galaxy might be a million light-years away. A light beam leaving that galaxy now will take a million years to make its trip to Earth; the light reaching our eyes now left that galaxy a million years ago. The more distant the objects that astronomers look at, the farther back they are looking in time.

The universe's fate hinges on how well our space-time balloon is expanding. If the expansion is slowing down rapidly, then it's a good sign that the energy from the big bang is nearly spent; our universe would be heading for a big crunch. On the other hand, if the universe's expansion isn't slowing down very much, the energy of the big bang might have given the fabric of space-time enough of a kick to let it expand for eternity.

Astronomers have begun to measure the change in the universe's expansion. A certain type of supernova (exploding star), called a type Ia, is a standard candle like Hubble's Cepheid stars. The Ia supernovas explode in roughly the same way and with the same brightness. But unlike Hubble's dim Cepheids, supernovas are visible halfway across the universe.

In late 1997 astronomers announced that they had used these supernovas to measure the distance to some very dim and ancient galaxies. The distance of the galaxy yields its age—and its Doppler shift yields its velocity. By comparing how fast galaxies were receding at different eras in the past, the astronomers were able to track how fast space-time was expanding. The answer they got was an odd one.

The expansion of the universe isn't slowing down. It might even be speeding up. The supernova data imply that the universe is getting bigger and bigger, faster and faster. If this is the case, there is little chance of a big crunch, because something is opposing the force of gravity. Once again physicists are talking about the cosmological constant—the mysterious term that Einstein added to his equations to balance the push of gravity. Einstein's biggest blunder might not have been a blunder after all.

The mysterious force, once again, might be the force of the vacuum. The tiny particles that seethe through space-time exert a gentle outward push, stretching the fabric of space-time imperceptibly. Over billions of years, that stretch adds up, and the universe inflates faster and faster. The fate of our universe will not be a big crunch but an eternal expansion, cooling, and heat death, thanks to the zero-point energy, a zero in the equations of quantum mechanics that imbues the vacuum with an infinity of particles.

Astronomers are still cautious. These supernova results are preliminary, but they are getting more solid with each observation. Other studies, which analyze plumes of gas or the number of *gravitational lenses* in a given field of view, also support the supernova results, implying that the cosmos will expand forever. The universe will die a cold death, not a hot one.

The answer is ice, not fire, thanks to the power of zero.

To Infinity and Beyond

However, if we do discover a complete theory, it should in time be understandable in broad principle by everyone, not just a few scientists. Then we shall all, philosophers, scientists, and just ordinary people, be able to take part in the discussion of the question of why it is that we and the universe exist. If we find the answer to that, it would be the ultimate triumph of human reason—for we would know the mind of God.

— STEPHEN HAWKING

Zero is behind all of the big puzzles in physics. The infinite density of the black hole is a division by zero. The big bang creation from the void is a division by zero. The infinite energy of the vacuum is a division by zero. Yet dividing by zero destroys the fabric of mathematics and the framework of logic—and threatens to undermine the very basis of science.

In Pythagoras's day, before the age of zero, pure logic reigned supreme. The universe was predictable and orderly. It was built upon rational numbers and implied the existence of God. Zeno's troubling paradox was explained away by banishing infinity and zero from the realm of numbers.

With the scientific revolution, the purely logical world gave way to an empirical one, based upon observation rather than philosophy. For Newton to explain the laws of the universe, he had to ignore the illogic within his calculus—an illogic caused by a division by zero.

Just as mathematicians and physicists managed to overcome the divison-by-zero problem in the calculus and set it once more upon a logical framework, zero returned in the equations of quantum mechanics and general relativity and, once again, tainted science with the infinite. At the zeros of the universe, logic fails. Quantum theory and relativity fall apart. To solve the problem, scientists set out to banish zero yet once more and unify the rules that govern the cosmos.

If scientists succeed, they will understand the laws of the universe. We would know the physical laws that dictate everything to the edges of space and time, from the beginning of the cosmos to its end. Humans would understand the cosmic whim that created the big bang. We would know the mind of God. But this time, zero might not be so easy to defeat.

The theories that unify quantum mechanics and general relativity, that describe the centers of black holes and explain the singularity of the big bang, are so far removed from experiment that it might be impossible to determine which are correct and which are not. The arguments of string theorists and cosmologists might be mathematically precise and at the same time be as useless as the philosophy of Pythagoras. Their mathematical theories might be beautiful and consistent and might seem to explain the nature of the universe — and be utterly wrong.

All that scientists know is the cosmos was spawned from nothing, and will return to the nothing from whence it came.

The universe begins and ends with zero.

Appendix **A**
Animal, Vegetable, or Minister?

Let a and b each be equal to 1. Since a and b are equal,

$$b^2 = ab \qquad\qquad \text{(eq. 1)}$$

Since a equals itself, it is obvious that

$$a^2 = a^2 \qquad\qquad \text{(eq. 2)}$$

Subtract equation 1 from equation 2. This yields

$$a^2 - b^2 = a^2 - ab \qquad\qquad \text{(eq. 3)}$$

We can factor both sides of the equation; $a^2 - ab$ equals $a(a-b)$. Likewise, $a^2 - b^2$ equals $(a+b)(a-b)$. (Nothing fishy is going on here. This statement is perfectly true. Plug in numbers and see for yourself!) Substituting into equation 3, we get

$$(a+b)(a-b) = a(a-b) \qquad\qquad \text{(eq. 4)}$$

So far, so good. Now divide both sides of the equation by $(a-b)$ and we get

$$a + b = a \qquad \text{(eq. 5)}$$

Subtract *a* from both sides and we get

$$b = 0 \qquad \text{(eq. 6)}$$

But we set *b* to 1 at the very beginning of this proof, so this means that

$$1 = 0 \qquad \text{(eq. 7)}$$

This is an important result. Going further, we know that Winston Churchill has one head. But one equals zero by equation 7, so that means that Winston has no head. Likewise, Churchill has zero leafy tops, therefore he has one leafy top. Multiplying both sides of equation 7 by 2, we see that

$$2 = 0 \qquad \text{(eq. 8)}$$

Churchill has two legs, therefore he has no legs. Churchill has two arms, therefore he has no arms. Now multiply equation 7 by Winston Churchill's waist size in inches. This means that

$$(\text{Winston's waist size}) = 0 \qquad \text{(eq. 9)}$$

This means that Winston Churchill tapers to a point. Now, what color is Winston Churchill? Take any beam of light that comes from him and select a photon. Multiply equation 7 by the wavelength, and we see that

$$(\text{Winston's photon's wavelength}) = 0 \qquad \text{(eq. 10)}$$

But multiplying equation 7 by 640 nanometers, we see that

$$640 = 0 \qquad \text{(eq. 11)}$$

Combining equations 10 and 11, we see that

$$(\text{Winston's photon's wavelength}) = 640 \text{ nanometers}$$

This means that this photon — or any other photon that comes from Mr. Churchill — is orange. Therefore Winston Churchill is a bright shade of orange.

To sum up, we have proved, mathematically, that Winston Churchill has no arms and no legs; instead of a head, he has a leafy top; he tapers to a point; and he is bright orange. Clearly, Winston Churchill is a carrot. (There is a simpler way to prove this. Adding 1 to both sides of equation 7 gives the equation

$$2 = 1$$

Winston Churchill and a carrot are two different things, therefore they are one thing. But that's not nearly as satisfying.)

What is wrong with this proof? There is only one step that is flawed, and that is the one where we go from equation 4 to equation 5. We divide by $a - b$. But look out. Since a and b are both equal to 1, $a - b = 1 - 1 = 0$. We have divided by zero, and we get the ridiculous statement that $1 = 0$. From there we can prove any statement in the universe, whether it is true or false. The whole framework of mathematics has exploded in our faces.

Used unwisely, zero has the power to destroy logic.

Appendix **B**
The Golden Ratio

Divide a line in two, such that the ratio of the small part to the large part is equal to the ratio of the large part to the whole line. For the sake of simplicity, let's say that the small part is 1 foot long.

If the small part is 1 foot long, and the large part is x feet long, then the length of the whole line is obviously $1 + x$ feet long. Putting our relationship into algebra, we find that the ratio of the small part to the large part is

$$1/x$$

while the ratio of the large part to the whole thing is

$$x/(1 + x)$$

Since the ratio of the small to the large is equal to the ratio of the large to the whole, we can set the two ratios equal to each other, yielding the equation

$$x/(1 + x) = 1/x$$

We wish to solve the equation for x, which is the golden ratio. The first step is to multiply both sides by x, which yields

$$x^2/(1 + x) = 1$$

We then multiply by $(1 + x)$, which gives us the equation

$$x^2 = 1 + x$$

Subtract $1 + x$ from both sides, yielding

$$x^2 - x - 1 = 0$$

Now we can solve the quadratic equation. This gives us two solutions:

$$(1 + \sqrt{5})/2 \text{ and } (1 - \sqrt{5})/2$$

Only the first one, with a value of about 1.618, is positive, thus it is the only one that made sense to the Greeks. Thus the golden ratio is approximately 1.618.

Appendix **C**
The Modern Definition
of a Derivative

Nowadays, the derivative is on firm logical grounds, because we define it in terms of limits. The formal definition of a derivative of a function $f(x)$ [which we denote as $f'(x)$] is

$$f'(x) = \underset{\text{(as } \varepsilon \text{ goes to 0)}}{\text{limit}} \quad \text{of } \frac{[f(x + \varepsilon) - f(x)]}{\varepsilon}$$

To see how this gets rid of Newton's dirty trick, let's look at the same function we used for demonstrating Newton's fluxions: $f'(x) = x^2 + x + 1$. The derivative of this function is equal to

$$f'(x) = \underset{\text{(as } \varepsilon \text{ goes to 0)}}{\text{limit}} \quad \text{of } \frac{[(x + \varepsilon)^2 + x + \varepsilon + 1 - (x^2 + x + 1)]}{\varepsilon}$$

Multiplying out, we get

$$f'(x) = \underset{\text{(as } \varepsilon \text{ goes to 0)}}{\text{limit}} \quad \text{of } \frac{[x^2 + 2\varepsilon x + \varepsilon^2 + x + \varepsilon + 1 - x^2 - x - 1]}{\varepsilon}$$

Now the x^2 cancels out the $-x^2$, the x cancels out the $-x$, and the 1 cancels the -1, leaving us with

$$f'(x) = \underset{\text{(as } \varepsilon \text{ goes to 0)}}{\text{limit}} \text{ of } \frac{[2\varepsilon x + \varepsilon + \varepsilon^2]}{\varepsilon}$$

Dividing through by ε—remember, ε is always nonzero because we haven't taken the limit yet—we get

$$f'(x) = \underset{\text{(as } \varepsilon \text{ goes to 0)}}{\text{limit}} \text{ of } 2x + 1 + \varepsilon$$

Now we take the limit and let ε approach zero. We get

$$f'(x) = 2x + 1 + 0 = 2x + 1$$

which is the answer we desired.

It's a small shift in thinking, but it makes all the difference in the world.

Appendix **D**
Cantor Enumerates the Rational Numbers

To show that the rational numbers are the same size as the natural numbers, all Cantor had to do was come up with a clever seating pattern. This is precisely what he did.

As you may recall, the rationals are the set of numbers that can be expressed as a/b for some integers a and b (with b nonzero, of course). To start with, let us consider the positive rational numbers.

Imagine a grid of numbers — two number lines crossed at zero, just like the Cartesian coordinate system. Let's put zero at the origin, and at every other point on that grid, let's associate with the rational number x/y where x is the point's x-coordinate and y is the point's y-coordinate. Since the number lines go off to infinity, every positive combination of xs and ys has a spot on that grid (Figure 58).

Now let's create a seating chart for the positive rational numbers. For seat one, let's start at 0 on the grid. Then let's move to 1/1: that's seat two. Next, wander over to 1/2: seat three. Then to 2/1 (which, of course, is the same thing as 2):

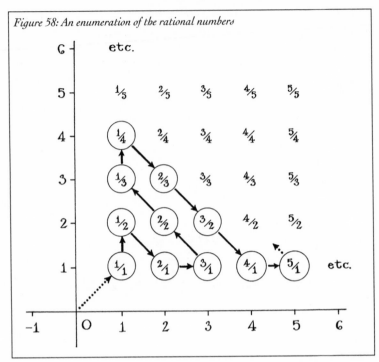

Figure 58: An enumeration of the rational numbers

seat four. Then to 3/1: seat five. We can wander back and forth along the grid, counting off numbers as we go. This yields a seating chart:

Seat	Rational number
1	0
2	1
3	1/2
4	2
5	3
6	1
7	1/3
8	1/4
9	2/3
etc.	etc.

Eventually, all the numbers have a seat; some, in fact, have two seats. It's easy to remove the duplicates—just skip over them when making the seating chart. The next step is to double the list, adding the negative rationals after the corresponding positive rational. This gives us a seating chart:

Seat	Rational number
1	0
2	1
3	−1
4	1/2
5	−1/2
6	2
7	−2
8	3
9	−3
etc.	etc.

Now all the rational numbers—positives, negatives, and zero—have a seat. Since nobody is left standing and no seats are open, this means that the rationals are the same size as the counting numbers.

Appendix **E**
Make Your Own Wormhole Time Machine

It's easy — just follow these four simple steps.

Step 1: Build a small wormhole. Both ends will be in the same point in time:

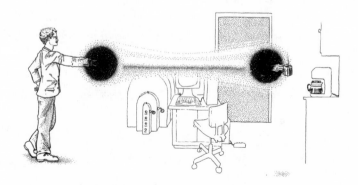

Step 2: Attach one end of the wormhole to something very heavy and the other end to a spaceship that's going at 90 percent of the speed of light. Every spaceship year is equivalent

to 2.3 years on Earth; clocks at either end of the wormhole will go at different speeds.

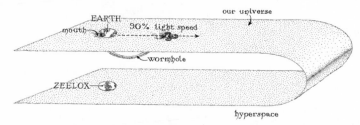

Step 3: Wait for a while. After 46 years of Earth time, bring the wormhole to a friendly planet. Traveling through the wormhole can take you from the year 2046 on Earth to the year 2020 on Zeelox or vice versa.

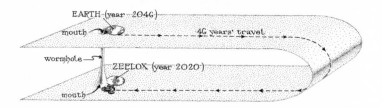

Step 4: If you were really smart, you could have started planning the mission far in advance. You could have sent a message to Zeelox long before you started, arranging for a spaceship from Zeelox to do the reverse process, beginning in 1974 (Zeelox time). Then in the year 2020 (Zeelox time), the other wormhole could transport you to Earth in the year 1994 (Earth time). If you use both wormholes, you can jump from 2046 (Earth time) to 2020 (Zeelox time) to 1994 (Earth time): you've traveled back in time more than half a century!

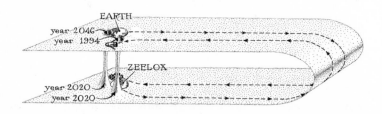

Selected Bibliography

Books and Periodicals

Aczel, Amir. *Fermat's Last Theorem.* New York: Delta, 1996.

Anselmo, Joseph. "Controller Error Lost Soho." *Aviation Week & Space Technology,* 7 July 1998: 28.

Aquinas, Thomas. *Selected Philosophical Writings.* Trans. Timothy McDermott. Oxford: Oxford University Press, 1993.

Aristophanes. *The Wasps / The Poet and the Women / The Frogs.* Trans. David Barrett. London: Penguin Books, 1964.

Aristotle. *The Metaphysics.* Trans. John McMahon. Amherst, N.Y.: Prometheus Books, 1991.

———. *Physics.* Trans. Robin Waterfield. New York: Oxford University Press, 1996.

Artin, E. *Modern Higher Algebra.* New York: New York University, 1947. (Lecture notes.)

St. Augustine. *Confessions.* Trans. Henry Chadwick. Oxford: Oxford University Press, 1991.

Beckmann, Petr. *A History of Pi.* New York: St. Martin's Press, 1971.

Bede. *Ecclesiastical History of the English People.* Trans. Leo Sherley-Price. London: Penguin Books, 1955.

Bell, E. T. *The Development of Mathematics.* New York: Dover Publications, 1940.

Berkovits, Nathan. "An Introduction to Superstring Theory and Its Duality Symmetries." Los Alamos National Labs Archive: hep-th/9707242.

Blay, Michel. *Reasoning with the Infinite.* Trans. M. B. DeBevoise. Chicago: University of Chicago Press, 1993.

Boethius. *The Consolation of Philosophy.* Trans. V. E. Watts. London: Penguin Books, 1969.

Book of the Dead. Trans. R. O. Faulkner. London: British Museum Publications, 1972.

Boyer, Carl B. *A History of Mathematics.* New York: John Wiley and Sons, 1968.

Bradford, Ernle. *The Sword and the Scimitar.* Milan: G. P. Putnam's Sons, 1974.

Bressoud, David. *Factorization and Primality Testing.* New York: Springer-Verlag, 1989.

Browne, Malcolm. "A Bet on a Cosmic Scale, and a Concession, Sort of." *New York Times,* 12 February 1997: A1.

Bunt, Lucas, Phillip Jones, and Jack Bedient. *The Historical Roots of Elementary Mathematics.* New York: Dover Publications, 1976.

Cantor, Georg. *Contributions to the Founding of the Theory of Transfinite Numbers.* Trans. Philip E. B. Jourdain. New York: Dover Publications, 1955.

Churchill, Ruel, and James Brown. *Complex Variables and Applications.* New York: McGraw-Hill, 1984.

Cipra, Barry. "In Mao's China, Politically Correct Math." *Science,* 28 February 1997: 1264.

Closs, Michael, ed. *Native American Mathematics.* Austin, Tex.: University of Texas Press, 1986.

Conway, John H. and Richard Guy. *The Book of Numbers.* New York: Copernicus, 1996.

Copleston, Frederick. *A History of Philosophy.* New York: Image Books, 1994.

Danzig, Tobias. *Number: The Language of Science.* New York: The Free Press, 1930.

David, Florence Nightingale. *Games, Gods and Gambling.* New York: Dover Publications, 1962, 1998.

Davies, Paul. "Paradox Lost." *New Scientist,* 21 March 1998: 26.

Dawson, John. *Logical Dilemmas.* Wellesley, Mass.: A. K. Peters, 1997.

Descartes, René. *Discourse on Method and Meditations on First Philosophy.* Ed. David Weissman. New Haven, Conn.: Yale University Press, 1996.

Diodorus Siculus. *Historical Library.* In *Diodorus of Sicily in Twelve Volumes with an English Translation* by C. H. Oldfather. Cambridge, Mass.: Harvard University Press; London: William Heinemann, Ltd., 1989, 1976.

Diogenes Laertius. *Lives of the Philosophers.* Trans. A. Robert Caponigri. Chicago: Henry Regnery Company, 1969.

———. *Lives of Eminent Philosophers.* Vols. 1–2. Trans. R. D. Hicks. Cambridge, Mass.: Harvard University Press, 1972.

DoCarmo, Manfredo. *Differential Geometry of Curves and Surfaces.* Englewood Cliffs, N.J.: Prentice-Hall, 1976.

Dubrovin, B. A., A. T. Fomenko, and S. P. Novikov. *Modern Geometry — Methods and Applications.* New York: Springer-Verlag, 1984.

Duhem, Pierre. *Medieval Cosmology*. Trans. Roger Ariew. Chicago: The University of Chicago Press, 1985.

Dym, H., and H. P. McKean. *Fourier Series and Integrals*. San Diego: Academic Press, 1972.

"Ebola: Ancient History of 'New' Disease?" *Science*, 14 June 1996: 1591.

Einstein, Albert. *Relativity*. New York: Crown Publishers, 1961.

Ekeland, Ivar. *The Broken Dice*. Trans. Carol Volk. Chicago: University of Chicago Press, 1991.

The Epic of Gilgamesh. Trans. N. K. Sanders. London: Penguin Books, 1960.

Feller, William. *An Introduction to Probability Theory and Its Applications*. New York: John Wiley and Sons, 1950.

Feynman, Richard. *QED*. Princeton, N.J.: Princeton University Press, 1985.

Feynman, Richard, et al. *The Feynman Lectures on Physics*. Reading, Mass.: Addison-Wesley, 1963.

Folland, Gerald. *Real Analysis*. New York: John Wiley and Sons, 1984.

Fulton, William. *Algebraic Curves*. Redwood City, Calif.: Addison-Wesley, 1969.

Fritzsch, Harald. *Quarks*. New York: Basic Books, 1983.

Garber, Daniel. *Descartes' Metaphysical Physics*. Chicago: University of Chicago Press, 1992.

Gaukroger, Stephen. *Descartes: An Intellectual Biography*. Oxford: Clarendon Press, 1995.

Gerdes, Paulus. *Marx Demystifies Calculus*. Trans. Beatrice Lumpkin. Minneapolis: MEP Publications, 1983.

Gödel, Kurt. *On Formally Undecidable Propositions of Principia Mathematica and Related Systems*. Trans. B. Meltzer. New York: Dover Publications, 1992.

Gould, Stephen Jay. *Questioning the Millennium*. New York: Harmony Books, 1997.

Graves, Robert. *The Greek Myths*. Vols. 1–2. London: Penguin Books, 1955.

Graves, Robert, and Raphael Patai. *Hebrew Myths: The Book of Genesis*. New York: Greenwich House, 1963.

Grun, Bernard. *The Timetables of History*. New York: Simon and Schuster, 1979.

Guthrie, Kenneth Sylvan. *The Pythagorean Sourcebook and Library*. Grand Rapids, Mich.: Phanes Press, 1987.

Hawking, Stephen. *A Brief History of Time*. New York: Bantam Books, 1988.

Hawking, S. W., and Neil Turok. "Open Inflation Without False Vacua." Los Alamos National Labs Archive: hep-th/9802030.

Hayashi, Alden. "Rough Sailing for Smart Ships." *Scientific American*, November 1988: 46.

Heath, Thomas. *A History of Greek Mathematics*. Vols. 1–2. New York: Dover Publications, 1921, 1981.

Heisenberg, Werner. *Physics and Philosophy: The Revolution in Modern Science.* New York: Harper and Row, 1958.

Hellemans, Alexander, and Bryan Bunch. *The Timetables of Science.* New York: Simon and Schuster, 1988.

Herodotus. *The Histories.* Trans. Aubrey de Selincourt. London: Penguin Books, 1954.

Hesiod. *Theogony.* In *The Homeric Hymns and Homerica with an English Translation* by Hugh G. Evelyn-White. Cambridge, Mass.: Harvard University Press; London: William Heinemann, Ltd., 1914. (Machine readable text.)

Hoffman, Paul. *The Man Who Loved Only Numbers.* New York: Hyperion, 1998.

———. "The Man Who Loves Only Numbers." *The Atlantic Monthly,* November 1987: 60.

Hooper, Alfred. *Makers of Mathematics.* New York: Random House, 1948.

Horgan, John. *The End of Science.* Reading, Mass.: Addison-Wesley, 1996.

Hungerford, Thomas. *Algebra.* New York: Springer-Verlag, 1974.

Iamblicus. *De Vita Pythagorica.* (On the Pythagoraean way of life). Trans. John Dillon, Jackson Hershbell. Atlanta: Scholars Press, 1991.

Jacobson, Nathan. *Basic Algebra I.* New York: W. H. Freeman and Company, 1985.

Jammer, Max. *Concepts of Space.* New York: Dover Publications, 1993.

Johnson, George. "Physical Laws Collide in a Black Hole Bet." *New York Times,* 7 April 1998: C1.

———. "Useful Invention or Absolute Truth: What Is Math?" *New York Times,* 10 February 1998: C1.

Kak, Subhash. "Early Theories on the Distance to the Sun." Los Alamos National Labs Archive: physics/9804021.

———. "The Speed of Light and Puranic Cosmology." Los Alamos National Labs Archive: physics/9804020.

Katz, Victor. *A History of Mathematics.* New York: HarperCollins College Publishers, 1993.

Kelly, Douglas. *Introduction to Probability.* New York: Macmillan Publishing Company, 1994.

Kestenbaum, David. "Practical Tests for an 'Untestable' Theory of Everything?" *Science,* 7 August 1998: 758.

Khayyám, Omar. *Rubaiyat of Omar Khayyám.* Trans. Edward Fitzgerald. Boston: Branden Publishing, 1992.

Knopp, Konrad. *Elements of the Theory of Functions.* Trans. Frederick Bagemihl. New York: Dover Publications, 1952.

———. *Theory of Functions.* Trans. Frederick Bagemihl. New York: Dover Publications, 1947, 1975.

Koestler, Arthur. *The Watershed.* New York: Anchor Books, 1960.

The Koran. Trans. N. J. Dawood. London: Penguin Books, 1936.

Lang, Serge. *Complex Analysis.* New York: Springer-Verlag, 1993.

———. *Undergraduate Algebra.* New York: Springer-Verlag, 1987.

Leibniz, Gottfried. *Discourse on Metaphysics / Correspondence with Arnauld / Monadology.* Trans. George Montgomery. La Salle, Ill.: Open Court Publishing, 1902.

Lo, K.Y. et al. "Intrinsic Size of SGR A*: 72 Schwarzschild Radii." Los Alamos National Labs Archive: astro-ph/9809222.

Maimonides, Moses. *The Guide for the Perplexed.* Trans. M. Friedlander. New York: Dover Publications, 1956.

Manchester, William. *A World Lit Only By Fire.* Boston: Little, Brown and Company, 1993.

Maor, Eli. *e: the Story of a Number.* Princeton, N.J.: Princeton University Press, 1994.

———. *To Infinity and Beyond.* Princeton, N.J.: Princeton University Press, 1987.

Marius, Richard. *Martin Luther.* Cambridge, Mass.: Belknap Press, 1999.

Matt, Daniel C. *The Essential Kabbalah.* San Francisco: Harper SanFrancisco, 1996.

Morris, Richard. *Achilles in the Quantum Universe.* New York: Henry Holt and Company, 1997.

Murray, Oswyn. *Early Greece.* Stanford, Calif.: Stanford University Press, 1980.

Nagel, Ernest and James Newman. *Gödel's Proof.* New York: New York University Press, 1958.

Neugebauer, O. *The Exact Sciences in Antiquity.* New York: Dover Publications, 1969.

Newman, James R. *The World of Mathematics.* Vols. 1–4. New York: Simon and Schuster, 1956.

Oldach, David, et al. "A Mysterious Death." *The New England Journal of Medicine.* 338 (11 June 1998): 1764.

Ovid. *Metamorphoses.* Trans. Rolfe Humphries. Bloomington, Ind.: Indiana University Press, 1955.

Oxtoby, David, and Norman Nachtrieb. *Principles of Modern Chemistry.* Philadelphia: Saunders College Publishing, 1986.

Pais, Abraham. *Subtle Is the Lord.* Oxford: Oxford University Press, 1982.

Pappas, Theoni. *Mathematical Scandals.* San Carlos, Calif.: Wide World Publishing/Tetra, 1997.

Pascal, Blaise. *Pensées and Other Writings.* Trans. Honor Levi. Oxford: Oxford University Press, 1995.

Pausanius. *Guide to Greece.* Vol. 2, *Description of Greece.* Trans. Peter Levi. London: Penguin Books, 1971.

Perricone, Mike. "The Universe Lives On." *FermiNews*, 19 June 1998: 5.

Pickover, Clifford. *The Loom of God.* New York: Plenum Press, 1997.

Plato. *Parmenides.* From *Plato in Twelve Volumes,* Vol. 1 translated by Harold North Fowler; introduction by W. R. M. Lamb (1966); Vol. 3 translated by W. R. M. Lamb (1967); Vol. 4 translated by Harold North Fowler (1977); Vol. 9 translated by Harold N. Fowler (1925). Cambridge, Mass.: Harvard University Press; London: William Heinemann, Ltd., 1966, 1967, 1977. (Machine readable text.)

Plaut, Gunther. *The Torah: A Modern Commentary*. New York: The Union of American Hebrew Congregations, 1981.

Plutarch. *Makers of Rome*. Trans. Ian Scott-Kilvert. London: Penguin Books, 1965.

Plutarch on Sparta. Trans. Richard J. A. Talbert. London: Penguin Books, 1988.

The Poetic Edda. Trans. Lee Hollander. Austin, Tex.: University of Texas Press, 1962.

Polchinski, Joseph. "String Duality." *Reviews of Modern Physics*, October 1996: 1245.

Pullman, Bernard, ed. *The Emergence of Complexity in Mathematics, Physics, Chemistry, and Biology*. Vatican City: Pontificia Academia Scientarum, 1996.

Randel, Don, ed. *The New Harvard Dictionary of Music*. Cambridge, Mass.: Belknap Press, 1986.

"Random Samples." *Science*, 23 January 1998: 487.

Rensberger, Boyce. "2001: A Calendar Odyssey." *Washington Post*, 11 December 1996: H01.

Resnikoff, H. L. and R. O. Wells. *Mathematics in Civilization*. New York: Dover Publications, 1973.

The Rig Veda. Trans. Wendy O'Flaherty. London: Penguin Books, 1981.

Rigatelli, Laura Toti. *Evariste Galois*. Trans. John Denton. Basel: Birkhauser Verlag, 1996.

Rothstein, Edward. *Emblems of Mind*. New York: Times Books, 1995.

Rotman, Brian. *Signifying Nothing*. Stanford, Calif.: Stanford University Press, 1987.

Royden, H. L. *Real Analysis*. New York: Macmillan, 1988.

Rudin, Walter. *Principles of Mathematical Analysis*. New York: McGraw-Hill, 1964.

———. *Real and Complex Analysis*. New York: McGraw-Hill, 1987.

Russell, Bertrand. *The Philosophy of Leibniz*. London: Routledge, 1992.

———. *The Philosophy of Logical Atomism*. Chicago: Open Court Publishing, 1985.

Schecter, Bruce. *My Brain Is Open*. New York: Simon and Schuster, 1998.

Scholem, Gershom. *Kabbalah*. New York: Meridian, 1974.

Seife, Charles. "Ever Outward." *New Scientist*, 3 January 1998: 19.

———. "Final Summer." *New Scientist*, 25 July 1998: 40.

———. "Magic Roundabout." *New Scientist*, 8 November 1997: 5.

———. "Mathemagician." *The Sciences*, May/June 1994: 12.

———. "Music and mathematics and their surprisingly harmonious relationship." *The Philadelphia Inquirer*, 21 May 1995: H3.

———. "No Way Out." *New Scientist*, 5 September 1998: 20.

———. "Running on Empty." *New Scientist*, 25 April 1998: 36.

———. "The Subtle Pull of Emptiness." *Science*, 10 January 1997: 158.

———. "Too Damned Hot." *New Scientist*, 1 August 1998: 21.

———. "Unlucky for Some." *New Scientist*, 4 July 1998: 23.

Singh, Simon. *Fermat's Enigma*. New York: Walker and Company, 1997.

Slakey, Francis. "Something Out of Nothing." *New Scientist*, 23 August 1997: 45.

Smith, David Eugene. *A Source Book in Mathematics*. New York: Dover Publications, 1959.

Sturluson, Snorri. *The Prose Edda*. Trans. Jean I. Young. Berkeley, Calif.: University of California Press, 1954.

Swetz, Frank. *From Five Fingers to Infinity*. Chicago: Open Court Publishing, 1994.

Teresi, Dick. "Zero." *The Atlantic Monthly*, July 1997: 88.

Thompson, Damian. *The End of Time*. Hanover, N.H.: University Press of New England, 1996.

Thorne, Kip. *Black Holes and Time Warps*. New York: W. W. Norton, 1994.

Thorpe, J. A. *Elementary Topics in Differential Geometry*. New York: Springer-Verlag, 1979.

Thucydides. *History of the Peloponnesian War*. Trans. Rex Warner. London: Penguin Books, 1954.

Turok, Neil, and S. W. Hawking. "Open Inflation, the Four Form and the Cosmological Constant." Los Alamos National Labs Archive: hep-th/9803156 v4.

The Upanishads. Trans. Juan Mascaro. London: Penguin Books, 1965.

Urmson, J. O., and Jonathan Ree, eds. *The Concise Encyclopedia of Western Philosophy and Philosophers*. London: Routledge, 1996.

Valenkin, Naum Yakovlevich. *In Search of Infinity*. Trans. Abe Shenitzer. Boston: Birkhauser, 1995.

Vilenkin, Alexander. "The Quantum Cosmology Debate." Los Alamos National Labs Archive: gr-qc/9812027.

Voltaire. *Candide*. London: Penguin Books, 1947.

Wang, Hao. *Reflections on Kurt Gödel*. Cambridge, Mass.: The MIT Press, 1987.

White, Michael. *The Last Sorcerer*. Reading, Mass.: Addison-Wesley, 1997.

Xenophon. *Hellenica*. Trans. Carleton L. Brownson. Vols. 1–2, *Xenophon in Seven Volumes*. Cambridge, Mass.: Harvard University Press; London: William Heinemann, Ltd.; Vol. 1, 1985; Vol. 2, 1986. (Machine readable text.)

Web Sites

Papyrus of Ani; Egyptian Book of the Dead. Trans. E. A. Wallis Budge. http://www.sas.upenn.edu/African_Studies/Books/Papyrus_Ani.html

Clement of Alexandria. *The Stromata*. http://www.webcom.com/~gnosis/library/strom4.htm

"The Life of Hypatia." http://www.cosmopolis.com/alexandria/

"Frequently Asked Questions in Mathematics." http://www.cs.unb.ca/~alopez-o/math-faq/

Odenwald, Sten. "Beyond the Big Bang." http://www2.ari.net/home/odenwald/anthol/beyondbb.html

————. "The Decay of the False Vacuum." http://www2.ari.net/home/odenwald/anthol/decay.html

Perseus Project. http://hydra.perseus.tufts.edu/

Slabodkin, Gregory. "Software glitches leave Navy Smart Ship dead in the water." *Government Computer News*, 13 July 1998. http://www.gcn.com/gcn/1998/July13/cov2.htm

The MacTutor History of Mathematics archive. http://history.math.csusb.edu/

Weisstein, Eric. Eric's Treasure Trove of Astronomy. http://www.treasure-troves.com/astro/astro0.html

Acknowledgments

A lot of people in my life are partly responsible, in one way or another, for this book. My high school teachers instilled a love of science and of writing; my college professors showed me the beauty of mathematics. Jeremy Bernstein and Kenneth Goldstein started my career of science journalism. My friends and colleagues at *New Scientist*—well, they put up with me, which is no mean feat.

Special thanks go to my agent, Kerry Nugent-Wells; the illustrator, Matt Zimet; and my editor, Wendy Wolf. Thanks, too, to the people who helped me whip the text into shape—Dawn Drzal, Faye Flam, and, of course, my mom and dad. (They're gentle parents, but harsh critics!)

Index

Page numbers in *italics* refer to illustrations.